How to Pass

SECOND EDITION

HIGHER

Maths

Brian J. Logan

HODDER
GIBSON
AN HACHETTE UK COMPANY

Although every effort has been made to ensure that website addresses are correct at time of going to press, Hodder Gibson cannot be held responsible for the content of any website mentioned in this book. It is sometimes possible to find a relocated web page by typing in the address of the home page for a website in the URL window of your browser.

Hachette UK's policy is to use papers that are natural, renewable and recyclable products and made from wood grown in well-managed forests and other controlled sources. The logging and manufacturing processes are expected to conform to the environmental regulations of the country of origin.

Orders: please contact Bookpoint Ltd, 130 Park Drive, Milton Park, Abingdon, Oxon OX14 4SE. Telephone: (44) 01235 827827. Fax: (44) 01235 400454. Email education@bookpoint.co.uk Lines are open from 9 a.m. to 5 p.m., Monday to Friday, with a 24-hour message answering service. Visit our website at www.hoddereducation.co.uk. If you have queries or questions that aren't about an order, you can contact us at hoddergibson@hodder.co.uk

© Brian J. Logan 2019

First published in 2019 by
Hodder Gibson, an imprint of Hodder Education
An Hachette UK Company
211 St Vincent Street
Glasgow, G2 5QY

Impression number	6	5	4	3
Year	2023	2022	2021	2020

Cover photo © imaginando - stock.adobe.com
Illustrations by Aptara, Inc.
Typeset in CronosPro-Lt 13/15 pt by Aptara, Inc.
Printed in India
A catalogue record for this title is available from the British Library.
ISBN: 978 1 5104 5227 5

MIX
Paper from
responsible sources
FSC™ C104740

SCOTLAND EXCEL

We are an approved supplier on the Scotland Excel framework.

Schools can find us on their procurement system as:

Hodder & Stoughton Limited t/a Hodder Gibson.

Contents

Introduction   iv

Chapter 1   Revision   1

Part 1 Algebra

Chapter 2   Polynomials   5

Chapter 3   Functions   12

Chapter 4   Quadratic theory   17

Chapter 5   Recurrence relations   23

Chapter 6   Logarithms   27

Part 2 Geometry

Chapter 7   The straight line   34

Chapter 8   The circle   41

Chapter 9   Vectors   46

Part 3 Trigonometry

Chapter 10   Basic trigonometry   53

Chapter 11   Addition and double-angle formulae   58

Chapter 12   The wave function   62

Part 4 Calculus

Chapter 13   Differentiation   67

Chapter 14   Integration   78

Chapter 15   Further calculus   84

Appendix 1 Practice Paper 1   89

Appendix 2 Practice Paper 2   95

Appendix 3 Solutions to *For practice* questions   102

Welcome to *How to Pass Higher Maths*

The intention of this book is to help you to achieve your aims in Higher Maths, whether simply to pass the exam or to achieve a top grade. The fact that you are reading this book is a very encouraging sign as it shows that you really want to do well and are prepared to do extra work in order to fulfil your ambition.

In this book you will find the following information about Higher Mathematics:

- an outline of the course and assessment
- worked examples on the type of questions likely to appear in your assessments
- helpful hints and tips on how to tackle questions
- information on how to avoid common errors
- a revision chapter on some of the key information from National 5 Mathematics
- suggestions for revising and studying effectively
- examples for you to practise at the end of each topic
- specimen examination papers with worked solutions.

The Higher Mathematics course

As you work through the course you will study algebra, geometry and trigonometry, as well as calculus. Calculus is a new topic which was not part of the National 5 syllabus.

It is important that you retain the skills you learned in National 5.

All the skills listed below will be covered in detail as you work through the book.

Skills

Manipulating algebraic expressions

Manipulating trigonometric expressions

Identifying and sketching related functions

Determining composite and inverse functions

Solving algebraic equations

Solving trigonometric equations

Determining vector connections

Working with vectors

Differentiating functions

Using differentiation to investigate the nature and properties of functions

Integrating functions

Using integration to calculate definite integrals

Applying differential calculus

Applying integral calculus

Applying algebraic skills to rectilinear shapes

Applying algebraic skills to circles and graphs

Modelling situations using sequences

Prior learning

Students studying Higher Mathematics are expected to have already attained the required skills to pass National 5 Mathematics or equivalent.

Higher Mathematics assessment

The assessment for Higher Mathematics usually takes place in early May and consists of two question papers.

Paper 1 is a non-calculator paper. It lasts 1 hour 30 minutes and is worth 70 marks. Paper 2 lasts 1 hour 45 minutes and is worth 80 marks. Students may use a calculator in Paper 2.

Both question papers will assess operational and reasoning skills, and will contain both short answer and extended response questions. For each question paper, candidates will be provided with a separate answer booklet to record their answers.

Some students may not be proceeding to the final examinations, but instead will be entered and assessed for the stand alone unit assessments; it should be noted that these no longer form part of the exam. Most of the content in this guide is still relevant for such students, however any assessment practice in this book now follows the format of the exam rather than the unit assessments.

More information about the course and the assessment is available on the SQA website – www.sqa.org.uk

The list of formulae

When you sit your assessments you will be provided with a list of formulae. This is invaluable. Check the list on the following page to make sure you know what formulae are on it as you will need to memorise any other formulae required. During assessments you should refer to the list regularly. Even if you think you know all the formulae check the list to ensure you are accurate. Also take great care when copying formulae from the list as it is easy to make a transcription error.

Circle:

The equation $x^2 + y^2 + 2gx + 2fy + c = 0$ represents a circle centre $(-g, -f)$ and radius $\sqrt{g^2 + f^2 - c}$.

The equation $(x - a)^2 + (y - b)^2 = r^2$ represents a circle centre (a, b) and radius r.

Scalar Product: $\mathbf{a.b} = |\mathbf{a}||\mathbf{b}| \cos \theta$, where θ is the angle between $\mathbf{a}$ and $\mathbf{b}$.

or $\mathbf{a.b} = a_1 b_1 + a_2 b_2 + a_3 b_3$ where $\mathbf{a} = \begin{pmatrix} a_1 \\ a_2 \\ a_3 \end{pmatrix}$ and $\mathbf{b} = \begin{pmatrix} b_1 \\ b_2 \\ b_3 \end{pmatrix}$

Trigonometric formulae:

$$\sin(A \pm B) = \sin A \cos B \pm \cos A \sin B$$
$$\cos(A \pm B) = \cos A \cos B \mp \sin A \sin B$$
$$\sin 2A = 2 \sin A \cos A$$
$$\cos 2A = \cos^2 A - \sin^2 A$$
$$= 2 \cos^2 A - 1$$
$$= 1 - 2 \sin^2 A$$

Table of standard derivatives:

$f(x)$	$f'(x)$
$\sin ax$	$a \cos ax$
$\cos ax$	$-a \sin ax$

Table of standard integrals:

$f(x)$	$\int f(x)dx$
$\sin ax$	$-\dfrac{1}{a} \cos ax + C$
$\cos ax$	$\dfrac{1}{a} \sin ax + C$

The order of topics in this book

Centres in different regions may teach the course content in different orders. However, *all* the course content necessary for your assessments is included in this book and has been arranged in a logical order.

The content has been divided into four parts:

Algebra	Polynomials, Functions, Quadratic theory, Recurrence relations, Logarithms	Approx. 30–40% of the overall marks
Geometry	The straight line, The circle, Vectors	Approx. 15–35% of the overall marks
Trigonometry	Basic trigonometry, Addition and double-angle formulae, The wave function	Approx. 15–40% of the overall marks
Calculus	Differentiation, Integration, Further calculus	Approx. 10–25% of the overall marks

Some advice on studying

The fact that you are studying Higher Mathematics means that you are probably already experienced in studying and sitting examinations. Even so, many students could study in a more effective way. A good way to study mathematics is to practise examples having paper, pencil and calculator at hand. As you start to use this book, you should try the examples *before* looking at the solutions. If you can do the examples, that is great because they are all of a standard that could appear in an examination or assessment. If you are unsure about an example, check the solution carefully then practise some similar examples from a textbook until you are confident. If you are still unsure, seek advice and help from your teacher or another student.

Confidence is important when you are doing mathematics. You often hear sportsmen and women talking about how confident they are that they will do well in a competition. Well, confidence comes through success. If you start to get questions correct in mathematics it breeds confidence and belief, whereas if you regularly fail to get questions correct it can lead to negative thoughts. So concentrate on positive aspects of your work while trying to work on areas where there is room for improvement. And don't be too hard on yourself. Remember that you could pass the examination with a score of 50% so you are allowed to make mistakes. In fact, you do not need to be a brilliant mathematician to pass Higher Mathematics. It is invariably the students who work hardest who do best at examination time.

Of course a key to success is preparation. The external examinations at the end of the session are demanding because they cover work from throughout the course and are an unknown quantity. However, you can prepare by finding out what types of questions are most likely to come up in the examinations. Check out specimen papers and past papers. Ask your teacher who will be able to tell you the most important areas to focus on. I hope that you prepare well enough so that you can go into examination situations with confidence ready to show all you have learned. You are bound to be nervous at examination time, but if you have prepared properly you will be excited too, so good luck.

Revision

As mentioned in the introduction, students studying Higher Mathematics are expected to have already attained the required skills to pass National 5 Mathematics or equivalent. However, the fact that you have passed an exam at this level does not guarantee that you are expert in every part of the course. Yet it is a fact that expertise in certain areas of the National 5 course is a requirement for success in Higher Mathematics. This is not the place to re-study the National 5 course in detail. Some key parts of the course, for example the straight line, completing the square, the discriminant, basic trigonometry and vectors will continue into the Higher Mathematics course and will be revised in the appropriate chapters. A list of *some* other important topics is given below.

What you should know

You should know:
- ★ how to expand brackets and simplify algebraic expressions
- ★ how to factorise quadratic expressions
- ★ how to solve quadratic equations
- ★ how to solve simultaneous equations
- ★ how to add, subtract, multiply and divide algebraic fractions
- ★ the laws of indices
- ★ how to simplify expressions involving surds
- ★ the properties of similar shapes.

The above list shows some of the topics which you will have to use frequently as you work on the Higher Mathematics course. Three of the topics on the list – surds, indices and similar shapes – cause particular difficulty for many students at National 5, yet are important in the Higher Mathematics course.

Surds

You should be able to simplify expressions involving surds and to express a fraction with a rational denominator.

Example

1 Simplify $\left(\sqrt{2}+\sqrt{6}\right)^{2}$.

Solution

$$\left(\sqrt{2}+\sqrt{6}\right)^{2} = \left(\sqrt{2}+\sqrt{6}\right)\left(\sqrt{2}+\sqrt{6}\right) = 2+2\sqrt{12}+6$$
$$= 8+2\sqrt{12} = 8+2\sqrt{4\times3} = 8+4\sqrt{3}$$

2 Express $\dfrac{20}{\sqrt{5}}$ with a rational denominator.

Solution

$$\frac{20}{\sqrt{5}} = \frac{20}{\sqrt{5}} \times \frac{\sqrt{5}}{\sqrt{5}} = \frac{20\sqrt{5}}{5} = 4\sqrt{5}$$

Indices

The topic of indices will crop up often in the calculus part of the Higher Mathematics course. Make sure you know, and can use, the laws of indices.

Key points

The laws of indices

Rule	Illustration
1 $\quad a^p \times a^q = a^{p+q}$	$a^4 \times a^8 = a^{4+8} = a^{12}$
2 $\quad a^p \div a^q = a^{p-q}$	$b^{10} \div b^7 = b^{10-7} = b^3$
3 $\quad (a^p)^q = a^{pq}$	$(c^5)^2 = c^{5 \times 2} = c^{10}$
4 $\quad (ab)^n = a^n b^n$	$(de^3)^4 = d^4(e^3)^4 = d^4 e^{3 \times 4} = d^4 e^{12}$
5 $\quad a^0 = 1$	$7^0 = 1$
6 $\quad a^{-n} = \dfrac{1}{a^n}$	$f^{-2} = \dfrac{1}{f^2}$
7 $\quad a^{\frac{m}{n}} = \left(\sqrt[n]{a}\right)^m$	$g^{\frac{2}{3}} = \left(\sqrt[3]{g}\right)^2$

Example

1 Simplify $\dfrac{8a^{\frac{5}{2}}}{2a^{\frac{1}{2}} \times a}$

Solution

$$\frac{8a^{\frac{5}{2}}}{2a^{\frac{1}{2}} \times a} = \frac{8a^{\frac{5}{2}}}{2a^{\frac{1}{2}+1}} = \frac{8a^{\frac{5}{2}}}{2a^{\frac{3}{2}}} = 4a^{\frac{5}{2}-\frac{3}{2}} = 4a^{\frac{2}{2}} = 4a$$

2 Evaluate $8^{\frac{2}{3}}$

Solution

$$8^{\frac{2}{3}} = \left(\sqrt[3]{8}\right)^2 = 2^2 = 4$$

Similar shapes

Occasionally a problem involving similar shapes may arise within a problem in Higher Mathematics.

Two shapes are similar if:
- their corresponding sides are in the same ratio
- they are equiangular.

In the case of triangles, if one of these conditions is true, then the other will also be true.

Example

The shaded rectangle in the diagram below is drawn with one vertex at the origin and the opposite vertex lying at E on the line CD, where C is the point (0, 6) and D is the point (18, 0).

The other two vertices of the rectangle are the points P (0, p) and D (q, 0).

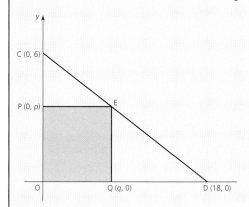

Show that $p = \frac{1}{3}(18 - q)$ and hence show that the area, $A(q)$, of the shaded rectangle is $A(q) = 6q - \frac{1}{3}q^2$.

Solution

Triangles OCD and QED are similar (because they are equiangular).

Hence $\dfrac{OD}{QD} = \dfrac{OC}{QE} \Rightarrow \dfrac{18}{18-q} = \dfrac{6}{p} \Rightarrow 18p = 6(18 - q)$

$$\Rightarrow p = \frac{6}{18}(18 - q) = \frac{1}{3}(18 - q).$$

Hence $A(q) = qp = q \times \frac{1}{3}(18 - q) = 6q - \frac{1}{3}q^2$.

It is worth noting that you should by now have memorised the formulae on the list of formulae in National 5 Mathematics. While these may not appear to be needed in your examinations at Higher Mathematics, it is as well to have them at your fingertips just in case.

Key points

1 The roots of $ax^2 + bx + c = 0$ are $x = \dfrac{-b \pm \sqrt{b^2 - 4ac}}{2a}$ (the quadratic formula)

2 Sine rule: $\dfrac{a}{\sin A} = \dfrac{b}{\sin B} = \dfrac{c}{\sin C}$

3 Cosine rule: $a^2 = b^2 + c^2 - 2bc \cos A$ or $\cos A = \dfrac{b^2 + c^2 - a^2}{2bc}$

4 Area of a triangle: $A = \dfrac{1}{2}bc \sin A$

The following section will allow you to practise some of the revision topics mentioned above. If you find you have difficulty with any of the following questions it would be best if you try to fix the problem either by asking a teacher or friend or by reference to a book on the National 5 Mathematics course.

For practice

1 Expand the brackets and simplify $(x + 2)(x^2 - 3x + 2)$.
2 Factorise $3x^2 - 6x - 24$.
3 Find the roots of the equation $3x^2 - 10x - 8 = 0$.
4 Solve the equation $x^2 - 4x - 2 = 0$, giving the roots correct to one decimal place.
5 Express $\dfrac{2}{x+3} - \dfrac{3}{x}$, $x \neq -3$, $x \neq 0$, as a single fraction in its simplest form.
6 Express $\sqrt{18} - \sqrt{2}$ as a surd in its simplest form.
7 Evaluate $16^{-\frac{3}{4}}$.

Polynomials

What you should know

You should know:
- ★ how to evaluate a polynomial expression
- ★ the remainder theorem
- ★ how to factorise a polynomial expression
- ★ how to solve a polynomial equation.

Examples of polynomial expressions are $f(x) = x^3 + 4x^2 - 5x + 2$ and $g(x) = 1 - 8x + 6x^3 - 2x^4$.

The polynomial $f(x) = x^3 + 4x^2 - 5x + 2$ is written in descending order.
The degree of the polynomial $f(x)$ is 3 as this is the highest power.
A polynomial of degree 3 is called a cubic polynomial.

The polynomial $g(x) = 1 - 8x + 6x^3 - 2x^4$ is written in ascending order.
The degree of the polynomial $g(x)$ is 4 as this is the highest power.
A polynomial of degree 4 is called a quartic polynomial.

Evaluating polynomials

Example

Evaluate $f(2)$ where $f(x) = x^3 + 4x^2 - 5x + 2$.

Solution

Either use substitution leading to $f(2) = 2^3 + 4 \times 2^2 - 5 \times 2 + 2 =$ $8 + 16 - 10 + 2 = 16$ or use the method of synthetic division in which the coefficients of the polynomial must be listed in *descending* order.

$$
\begin{array}{r|rrrr}
2 & 1 & 4 & -5 & 2 \\
  &   & 2 & 12 & 14 \\
\hline
  & 1 & 6 & 7 & \mathbf{16}
\end{array}
$$

The remainder theorem

1   The remainder theorem states that the remainder on dividing $f(x)$ by $(x - h)$ is $f(h)$.

Example

1 Find the remainder on dividing $1 - 8x + 6x^3 - 2x^4$ by $(x + 3)$.

Solution

Using the remainder theorem, the remainder is $f(-3)$. Now use synthetic division.

$$
\begin{array}{r|rrrrr}
-3 & -2 & 6 & 0 & -8 & 1 \\
   &   & 6 & -36 & 108 & -300 \\
\hline
   & -2 & 12 & -36 & 100 & -299
\end{array}
$$
$\Rightarrow$ remainder is -299

You may be asked to find the numerical value of a coefficient in a polynomial. This is more difficult involving reasoning skills.

2 Find the value of k if the remainder on dividing $x^3 + 2x^2 + kx - 9$ by $(x - 2)$ is 15.

Solution

Using the remainder theorem, the remainder is $f(2)$. Now use synthetic division.

$$
\begin{array}{r|rrrr}
2 & 1 & 2 & k & -9 \\
  &   & 2 & 8 & 2k + 16 \\
\hline
  & 1 & 4 & k + 8 & 2k + 7
\end{array}
$$
$\Rightarrow 2k + 7 = 15 \Rightarrow k = 4$

2   The remainder theorem can be used to show that the remainder on dividing $f(x)$ by $(ax - b)$ is $f\left(\dfrac{b}{a}\right)$.

Example

Find the remainder on dividing $2x^3 - 5x^2 + 6x - 3$ by $(2x - 1)$.

Solution

Using the remainder theorem, the remainder is $f\left(\dfrac{1}{2}\right)$. Now use synthetic division.

$$
\begin{array}{c|cccc}
\frac{1}{2} & 2 & -5 & 6 & -3 \\
 & & 1 & -2 & 2 \\
\hline
 & 2 & -4 & 4 & -1
\end{array}
\quad \Rightarrow \quad \text{the remainder is } -1
$$

The factor theorem

When a number or expression is divided by a factor, there is no remainder because a factor divides into a number or expression exactly. This leads to an extension of the remainder theorem called the factor theorem.

The factor theorem states that if $(x - a)$ is a factor of a polynomial $f(x)$ then $f(a) = 0$.

The theorem can be written the other way round, namely if $f(a) = 0$, then $(x - a)$ is a factor of a polynomial $f(x)$.

Example

Show that $(x - 3)$ is a factor of the polynomial $x^3 - 4x^2 + x + 6$.

Solution

Let $f(x) = x^3 - 4x^2 + x + 6$. Using the remainder theorem, the remainder when $f(x)$ is divided by $(x - 3)$ is $f(3)$. Now use synthetic division.

$$
\begin{array}{c|cccc}
3 & 1 & -4 & 1 & 6 \\
 & & 3 & -3 & -6 \\
\hline
 & 1 & -1 & -2 & 0
\end{array}
$$

Hence $(x - 3)$ is a factor $x^3 - 4x^2 + x + 6$.

Hints & tips

*When you carry out a synthetic division, the process gives you the **quotient** (the answer to the division) and the remainder (look along the last line). The coefficients of the terms in the quotient in the example above (1, −1 and −2) show that when $x^3 - 4x^2 + x + 6$ is divided by $(x - 3)$, then the quotient is $x^2 - x - 2$ and the remainder is 0.*

Factorising polynomials

We can use the ideas on the previous page to factorise polynomials. In the above example we could now factorise $x^3 - 4x^2 + x + 6$. We see that $x^3 - 4x^2 + x + 6 = (x - 3)(x^2 - x - 2)$. This can be factorised *fully* because $(x^2 - x - 2)$ is a quadratic expression which factorises further, leading to $x^3 - 4x^2 + x + 6 = (x - 3)(x - 2)(x + 1)$.

Hints & tips

Before we look at factorising polynomials in more detail, it should be pointed out that it is essential that you are comfortable with the factorisation of quadratic expressions. This features in several topics in the Higher Mathematics course. If you have difficulties in this area, you must take steps to remedy the situation.

Example

1 a) Show that $(3x - 2)$ is a factor of $3x^3 - 8x^2 + 7x - 2$.
  b) Factorise $3x^3 - 8x^2 + 7x - 2$ fully.

Solution

a) Let $f(x) = 3x^3 - 8x^2 + 7x - 2$. Using the remainder theorem, the remainder is $f\left(\frac{2}{3}\right)$. Now use synthetic division.

$$
\begin{array}{c|cccc}
\frac{2}{3} & 3 & -8 & 7 & -2 \\
 & & 2 & -4 & 2 \\
\hline
 & 3 & -6 & 3 & 0
\end{array}
$$

Hence $(3x - 2)$ is a factor of $3x^3 - 8x^2 + 7x - 2$.

b) $3x^3 - 8x^2 + 7x - 2 = \left(x - \frac{2}{3}\right)(3x^2 - 6x + 3)$

$$= \left(x - \frac{2}{3}\right) \times 3(x^2 - 2x + 1)$$

$$= (3x - 2)(x - 1)(x - 1)$$

Factorising the polynomial above was made simpler because we were given a start by being asked to show that $(3x - 2)$ was a factor. It becomes more difficult to factorise a polynomial when there is no helpful information to lead you into the example. We shall investigate this next.

2 Factorise $2x^3 + 5x^2 - 23x + 10$ fully.

Solution

Let $f(x) = 2x^3 + 5x^2 - 23x + 10$. To start, we must find a factor of $f(x)$ by trial. Use synthetic division trying whole numbers which are factors of 10. Start by trying 1, −1, 2, −2, 5, −5 and so on until a factor is identified.

```
1 |  2    5   -23   10
   |       2    7   -16
   |_____
      2    7   -16   -6  ≠ 0  ⇒  (x – 1) is not a factor of f(x)

-1 |  2    5   -23   10
   |      -2   -3    26
   |_____
      2    3   -26   36  ≠ 0  ⇒  (x + 1) is not a factor of f(x)

2 |  2    5   -23   10
   |       4   18  -10
   |_____
      2    9   -5    0  ⇒  (x – 2) is a factor of f(x)
```

Hence $2x^3 + 5x^2 - 23x + 10 = (x - 2)(2x^2 + 9x - 5)$
$$= (x - 2)(2x - 1)(x + 5).$$

Solving polynomial equations

Equations involving polynomials are often solved by factorising the polynomial.

Example

Solve the equation $2x^3 + 5x^2 - 23x + 10 = 0$.

Solution

We start by factorising $2x^3 + 5x^2 - 23x + 10$ (see previous example) then find the solution by equating each of the factors to zero.

Hence $2x^3 + 5x^2 - 23x + 10 = 0 \Rightarrow (x - 2)(2x - 1)(x + 5) = 0 \Rightarrow$
$x = 2, 0{\cdot}5, -5$.

Note that 2, $0{\cdot}5$ and -5 are called the roots of the equation. They indicate that the graph of the function $f(x) = 2x^3 + 5x^2 - 23x + 10$ crosses the x-axis at the points $(-5, 0)$, $(0{\cdot}5, 0)$ and $(2, 0)$.

Hints & tips

*Remember to state the roots only if you are asked to **solve** an equation, such as $2x^3 + 5x^2 - 23x + 10 = 0$. If you are only asked to factorise $2x^3 + 5x^2 - 23x + 10$, then it would be incorrect to include the roots.*

The graph of a polynomial function

Inspect the graph of the function $f(x) = 2x^3 + 5x^2 - 23x + 10$ shown.

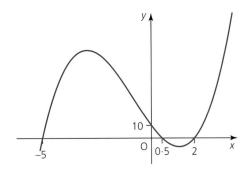

Note that the graph crosses the x-axis at the points $(-5, 0)$, $(0\cdot5, 0)$ and $(2, 0)$. It crosses the y-axis at the point $(0, 10)$. This can be found by replacing x by 0 to find $f(0) = 10$.

You should also be aware that there is a series of related polynomial graphs passing through $(-5, 0)$, $(0\cdot5, 0)$ and $(2, 0)$. They have equation $f(x) = k(2x^3 + 5x^2 - 23x + 10)$ or $f(x) = k(x - 2)(2x - 1)(x + 5)$ where k is a constant.

Finding the equation of a polynomial

Example

The graph of a cubic function crosses the x-axis at the points $(-2, 0)$, $(1, 0)$ and $(3, 0)$.

The graph crosses the y-axis at the point $(0, 24)$.

Find the equation of the cubic function.

Solution

The roots occur at -2, 1 and 3 hence $f(x) = k(x + 2)(x - 1)(x - 3)$.

Substitute $(0, 24)$ into this equation $\Rightarrow 24 = k \times 2 \times (-1) \times (-3) \Rightarrow 6k = 24 \Rightarrow k = 4$.

Hence the equation is $f(x) = 4(x + 2)(x - 1)(x - 3)$.

Hints & tips

Note that the above equation could be given in the form $f(x) = 4x^3 - 8x^2 - 20x + 24$ by expanding the brackets. It is not necessary to do this but check that you can expand the brackets accurately. It will be a good test of your algebra skills.

Finding points of intersection of two graphs

Example

Find the coordinates of the points where the line with equation $y = x - 2$ intersects the graph with equation $y = x^3 + x^2 - 4x + 1$.

Solution

Where the two graphs intersect, the y-coordinates will be the same.

Hence $x^3 + x^2 - 4x + 1 = x - 2 \Rightarrow x^3 + x^2 - 5x + 3 = 0$.

To factorise, try factors of 3, i.e. $\pm 1, \pm 3$.

$$
\begin{array}{r|rrrr}
1 & 1 & 1 & -5 & 3 \\
  &   & 1 & 2 & -3 \\
\hline
  & 1 & 2 & -3 & 0
\end{array}
$$

As the remainder = 0, $(x - 1)$ is a factor.

Hence $x^3 + x^2 - 5x + 3 = 0$

$\Rightarrow (x - 1)(x^2 + 2x - 3) = 0$

$\Rightarrow (x - 1)(x - 1)(x + 3) = 0$

$\Rightarrow x = 1$ or $x = -3$

Now substitute both values of x into $y = x - 2$ to find the y-coordinates of the points of intersection.

Hence the points of intersection are $(1, -1)$ and $(-3, -5)$.

For practice

1 **a)** Show that $(x + 1)$ is a factor of $x^3 - 3x^2 + 4$.
   b) Hence factorise $x^3 - 3x^2 + 4$ fully.
2 **a)** Show that $(x - 3)$ is a factor of $f(x) = 6x^3 - 25x^2 + x + 60$.
   b) Hence factorise $f(x)$ fully.
3 Given that $(x - 1)$ is a factor of $(x^3 - 5x^2 + kx - 5)$, find the value of k.
4 Given that $(3x - 1)$ is a factor of $(30x^3 + 17x^2 + kx - 2)$, find the value of k and then factorise the expression fully when k has this value.
5 **a)** Factorise fully $x^3 - 3x^2 - x + 3$.
   b) Hence solve the equation $x^3 - 3x^2 - x + 3 = 0$.

You should know:
- ★ the meaning of the domain and range of a function
- ★ how to find a formula for the composition of two functions, $f(g(x))$
- ★ what is meant by the inverse of a function
- ★ how to find a formula for an inverse function, $f^{-1}(x)$
- ★ the graphs of exponential and logarithmic functions
- ★ how, given the graph of a function, to sketch the graphs of related functions.

Domain and range

You should know that in mathematics, a *function* is a relationship between two sets such that each member of the first set is related to exactly one member of the second set. A simple example of a function is $f(x) = x^2$. By squaring members of the set of real numbers, R, we can find an image for each real number leading to pairs of numbers such as $(-5, 25)$, $(0, 0)$, $(4, 16)$ and so on. By plotting these points we can draw the graph of the function $y = x^2$.

The *domain* of a function is the set of values put into the function (the x values).

The *range* of a function is the set of values coming out of the function (the y values).

So for the function $f(x) = x^2$, we could say that -5 is a member of the domain and 25 is a member of the range and so on.

For the most part, the domain will be R, the set of real numbers.

But there are some exceptions which we will look at now. These often concern two 'impossible' calculations in mathematics, namely dividing by zero and finding the square root of a negative number.

Example

A function is given by $f(x) = \dfrac{5}{x-3}$. What is the largest possible domain on which f could be defined?

Solution

We can find an image for every real number using this function with one exception because division by zero is impossible. If $x = 3$, there would be no related member in the range, therefore the largest possible domain is $x \in R$, $x \neq 3$. This means that x is a member of the set of real numbers with the exception of 3.

Example

A function is given by $f(x) = \sqrt{x-8}$. What is the largest possible domain on which f could be defined?

Solution

Because we cannot find the square root of a negative number, $x - 8 \geq 0$. Therefore the largest possible domain is $x \in R, x \geq 8$.

Hints & tips

Check these two examples carefully and if you are asked to find the largest possible domain, think about the two 'impossible' calculations.

Composition of functions

A composition of functions occurs when we apply one function to the result of another. For example, suppose we have two functions given by $f(x) = 3x + 2$ and $g(x) = 7 - x$. We might be asked to calculate $f(g(3))$. This means we must apply function $f(x)$ to the result of $g(3)$. Hence $f(g(3)) = f(7 - 3) = f(4) = 3 \times 4 + 2 = 14$. We can find a formula for the composite function.

Example

If $f(x) = 3x + 2$ and $g(x) = 7 - x$, find a formula for

  a) $f(g(x))$ **b)** $g(f(x))$.

Solution

  a) $f(g(x)) = f(7 - x) = 3(7 - x) + 2 = 21 - 3x + 2 = 23 - 3x$

  b) $g(f(x)) = g(3x + 2) = 7 - (3x + 2) = 7 - 3x - 2 = 5 - 3x$

Hints & tips

Use the formula for $f(g(x))$ to confirm that $f(g(3)) = 14$. Note that the order of applying the functions is vital because $f(g(x)) \neq g(f(x))$ in this case. Remember that $f(g(x))$ means 'f after g'.

Example

If $f(x) = 2x + 1$ and $g(x) = \frac{1}{2}(x - 1)$, find a formula for $f(g(x))$.

Solution

$$f(g(x)) = f\left(\frac{1}{2}(x - 1)\right) = 2\left(\frac{1}{2}(x - 1)\right) + 1 = x - 1 + 1 = x$$

In this example, we find that $f(g(x)) = x$. No matter what value x is given its image will be the same. Therefore $f(x)$ and $g(x)$ are *inverses* of each other. This means that applying a function after its inverse gives us back the original value. The notation for the inverse of a function is $f^{-1}(x)$. Hence if $f(x) = 2x + 1$, then $f^{-1}(x) = \frac{1}{2}(x - 1)$. In $f(x)$ we multiply by 2 then add 1. In the inverse $f^{-1}(x)$ we undo this by subtracting 1 then dividing by 2.

Not every function has an inverse. An inverse only exists when there is a special type of function called a 1-1 correspondence. This is a function in which every member of the first set is paired with exactly one member of the second set and vice versa.

Hints & tips

If two functions $f(x)$ and $g(x)$ are such that $f(g(x)) = x$, then it tells you that $f(x)$ and $g(x)$ are inverse functions. This is a useful piece of information to remember.

Inverse functions

Given the formula for a function, we can find a formula for the inverse function by changing the subject of the formula.

Example

A function is given by $f(x) = 3x - 8$. Find the inverse function $f^{-1}(x)$.

Solution

Let $f(x) = y$, hence $y = 3x - 8$. Now change the subject of this formula to x.

$y = 3x - 8 \Rightarrow 3x = y + 8 \Rightarrow x = \dfrac{y+8}{3}$

Hence $f^{-1}(x) = \dfrac{x+8}{3}$ (replace y by x to complete the solution).

Hints & tips

You should check your solution by substituting a simple value for x in $f(x)$. For example, check that $f(5) = 7$ and then $f^{-1}(7) = 5$ to verify your solution.

The exponential and logarithmic functions

You will meet two important functions called the *exponential* function and the *logarithmic* function as you study Higher Mathematics.

The function $f(x) = a^x$ is called the exponential function.

The function $f(x) = \log_a x$ is called the logarithmic function.

These two functions are inverses of each other.

Their graphs are shown in the diagram on the following page.

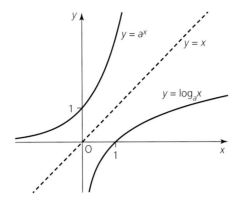

In these functions, *a* is called the *base* and *a* > 1.

The line *y* = *x* is the axis of symmetry in this diagram. In fact, if you are given the graph of *any* function *y* = *f*(*x*), then the graph of the inverse function *y* = *f*⁻¹(*x*) is the reflection of the graph in the line *y* = *x*.

The graphs of related functions

If you are given the graph of a function *y* = *f*(*x*), the following transformations describe how to sketch the graphs of related functions.

Key points

y = −*f*(*x*): reflect *y* = *f*(*x*) in the *x*-axis

y = *f*(−*x*): reflect *y* = *f*(*x*) in the *y*-axis

y = −*f*(−*x*): give *y* = *f*(*x*) a half turn rotation about the origin O

y = *f*(*x*) + *a*: slide *y* = *f*(*x*) *a* units up (parallel to the *y*-axis)

y = *f*(*x* − *a*): slide *y* = *f*(*x*) *a* units to the right (parallel to the *x*-axis)

y = *f*(*x* + *a*): slide *y* = *f*(*x*) *a* units to the left (parallel to the *x*-axis)

y = *af*(*x*): stretch *y* = *f*(*x*) parallel to the *y*-axis by a scale factor of *a*

$y = f\left(\dfrac{x}{a}\right)$: stretch *y* = *f*(*x*) parallel to the *x*-axis by a scale factor of *a*

y = *f*⁻¹(*x*): reflect *y* = *f*(*x*) in the line *y* = *x*.

You are likely to be asked to sketch the graph of a function which combines two of the above transformations.

Example

The diagram shows a sketch of the function *y* = *f*(*x*).

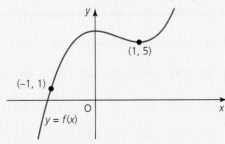

Sketch the graph of *y* = 3 − *f*(*x*).

Solution

To draw the graph of $y = 3 - f(x)$, think of it as $y = -f(x) + 3$, then reflect the graph of $y = f(x)$ in the x-axis, followed by moving it up 3 units in the direction of the y-axis. You should check that $(-1, 1) \rightarrow (-1, -1) \rightarrow (-1, 2)$ and $(1, 5) \rightarrow (1, -5) \rightarrow (1, -2)$.

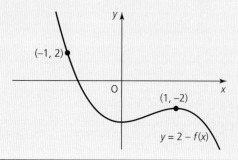

Hints & tips

In the above question, you can copy the original diagram, show the reflection of the original graph in the x-axis followed by the final graph after moving the reflection 3 units upwards. Always indicate your final solution clearly with a label or a key, or on a separate diagram. Always remember to label the x-axis and y-axis and to include the coordinates of the points used to find the solution.

For practice

1 A function is given by $f(x) = \dfrac{3}{(x+2)(x-4)}$. What is the largest possible domain on which f could be defined?

2 If $f(x) = 5x - 1$ and $g(x) = x^2 + 2$, find a formula for
   a) $f(g(x))$
   b) $g(f(x))$.

3 a) If $f(x) = 4x + 2$ and $g(x) = \dfrac{1}{4}(x-2)$, find a formula for $f(g(x))$.
   b) State the relationship between functions $f(x)$ and $g(x)$.

4 A function is given by $f(x) = 3 - 4x$. Find the inverse function $f^{-1}(x)$.

5 The diagram shows a sketch of the function $y = f(x)$.
   Sketch the graph of $y = f(x - 3) - 2$.

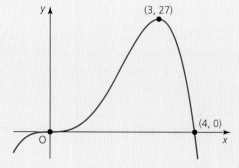

Chapter 4
Quadratic theory

What you should know

You should know:

★ how to solve quadratic equations by factorisation and by the quadratic formula
★ how to complete the square in a quadratic expression
★ how to sketch the graph of a quadratic function
★ how to find the equation of a quadratic function from its roots
★ how to use the discriminant to find the nature of the roots of a quadratic equation
★ how to solve a quadratic inequality by factorising
★ how to solve problems involving the intersection of a line and a parabola.

Solving quadratic equations

You should be able to solve quadratic equations by factorisation and by using the quadratic formula.

Example

Solve the equation $4x^2 + 2x - 2 = 0$.

Solution

$4x^2 + 2x - 2 = 0 \Rightarrow 2(2x^2 + x - 1) = 0 \Rightarrow 2(2x - 1)(x + 1) = 0 \Rightarrow x = \frac{1}{2}$ or

$x = -1$. You should remember the quadratic formula $x = \frac{-b \pm \sqrt{b^2 - 4ac}}{2a}$.

This can be used to solve any quadratic equation. Check that you can use this formula accurately by solving the equation $2x^2 - 5x - 4 = 0$ giving the roots correct to one decimal place. Well done if you got $x = 3 \cdot 1$ or $x = -0 \cdot 6$.

Completing the square

You should already know how to express a quadratic expression in the form $(x + p)^2 + q$. This process is known as *completing the square*.

Example

Express $x^2 + 6x + 2$ in the form $(x + p)^2 + q$.

Solution

$x^2 + 6x + 2 = x^2 + 6x + 9 - 9 + 2 = (x + 3)^2 - 7$.

It can be very helpful to use this form. If you are asked to sketch the graph of the parabola with equation $y = x^2 + 6x + 2$, then by expressing the equation in the form $y = (x + 3)^2 - 7$, you can tell that the parabola has a turning point at $(-3, 7)$. Note that the turning point will be a minimum as the coefficient of the x^2 term is positive.

Example

Express $2x^2 - 12x - 5$ in the form $a(x + p)^2 + q$.

Solution

$2x^2 - 12x - 5 = 2(x^2 - 6x) - 5 = 2(x^2 - 6x + 9) - 18 - 5 = 2(x - 3)^2 - 23$

The previous example was more difficult because the coefficient of the x^2 term was 2. To carry out the process, we take out 2 as a common factor of the first two terms. It is recommended that when you solve this type of problem, i.e. get $2(x - 3)^2 - 23$, you should multiply out the brackets to check it is correct. Note that the parabola with equation $y = 2x^2 - 12x - 5 = 2(x - 3)^2 - 23$ has a minimum turning point at $(3, -23)$.

The graph of a quadratic function

The method of completing the square is useful if you wish to make a quick sketch of the graph of a quadratic function. You will quickly spot the coordinates and nature of the turning point. It is also easy to see where the graph crosses the y-axis.

Example

Sketch the graph of the parabola with equation $y = 3 + 8x - x^2$.

Solution

$3 + 8x - x^2 = 3 - (x^2 - 8x) = 3 - (x^2 - 8x + 16) + 16 = 19 - (x - 4)^2$

Therefore the parabola with equation $y = 3 + 8x - x^2$ has a turning point at $(4, 19)$. It is a maximum turning point because the coefficient of the original x^2 term is negative. We can also see that the parabola

$\Rightarrow$

crosses the y-axis at $(0, 3)$. Why? Because if we replace x by 0 in the equation $y = 3 + 8x - x^2$, then $y = 3$.

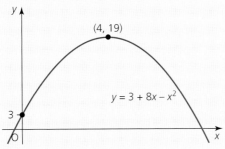

Finding the equation of a quadratic function from its roots

Example

What is the equation of the graph shown?

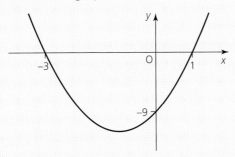

Solution

As the roots of the equation are -3 and 1, then $(x + 3)$ and $(x - 1)$ are factors of $f(x)$. We can find many graphs of the form $y = f(x)$ with these roots by stretching the graph shown parallel to the y-axis.

The equation must be in the form $y = k(x + 3)(x - 1)$. Use the point where the graph crosses the y-axis to find k. Substitute $x = 0$ and $y = -9$ into the equation $y = k(x + 3)(x - 1)$.

Hence $-9 = k(0 + 3)(0 - 1) \Rightarrow -3k = -9 \Rightarrow k = 3$.

Therefore the equation of the graph shown is $y = 3(x + 3)(x - 1)$ or $y = 3x^2 + 6x - 9$.

Hints & tips ★

Note that there is a neat way to find the coordinates of the turning point of the last graph. The axis of symmetry of the parabola lies midway between the roots (-3 and 1) and so has equation $x = -1$. By substituting $x = -1$ into the equation $y = 3(x + 3)(x - 1)$, check that $y = -12$. Hence the coordinates of the turning point of the graph are $(-1, -12)$.

The discriminant

You should already know that for the equation $ax^2 + bx + c = 0$, the *discriminant* is $b^2 - 4ac$, and that

- $b^2 - 4ac > 0 \Rightarrow$ two distinct real roots
- $b^2 - 4ac = 0 \Rightarrow$ equal real roots
- $b^2 - 4ac < 0 \Rightarrow$ no real roots.

When $b^2 - 4ac > 0$ and is also a perfect square, then the roots of the quadratic equation will be rational numbers. The expression $ax^2 + bx + c$ will factorise in this case.

When $b^2 - 4ac > 0$ and is not a perfect square, then the roots of the quadratic equation will be irrational numbers and the expression $ax^2 + bx + c$ will not factorise in this case.

When $b^2 - 4ac \geq 0$, we can say that real roots exist.

Example

Find the value of m such that the equation $mx^2 + (3m - 4)x + 4 = 0$ has equal roots.

Solution

The phrase 'has equal roots' gives a strong clue that this question involves the discriminant, so start by stating that $a = m$, $b = (3m - 4)$, $c = 4$.

So $b^2 - 4ac = (3m - 4)^2 - 4 \times m \times 4 = 9m^2 - 24m + 16 - 16m = 9m^2 - 40m + 16$.

As the condition for equal roots is $b^2 - 4ac = 0$ then $9m^2 - 40m + 16 = 0$.

Hence $(9m - 4)(m - 4) = 0 \Rightarrow m = \dfrac{4}{9}$ or 4.

Hints & tips

In order to solve this type of problem you had to be able to expand a bracket, factorise a quadratic expression, and simplify an algebraic expression. All these areas were covered before you started studying Higher Mathematics. Many students know how to do problems but lose marks through basic errors. So concentrate on eliminating this type of error. Remember also that questions involving the nature of the roots of a quadratic equation often involve the discriminant.

Quadratic inequalities

You should be able to solve a quadratic inequality. The following example will explain the steps required to carry out this process.

Example

Solve the inequality $15 - 2x - x^2 > 0$.

Solution

Start by finding the roots of the equation $15 - 2x - x^2 = 0$ using factorisation.

This leads to $15 - 2x - x^2 = 0 \Rightarrow (5 + x)(3 - x) = 0 \Rightarrow x = -5$ or 3.

Next sketch the graph of $y = 15 - 2x - x^2$. We know that the graph crosses the x-axis at -5 and 3. We can see the graph must have a maximum turning point as the x^2 term is negative. These are the *only* bits of information required for our sketch.

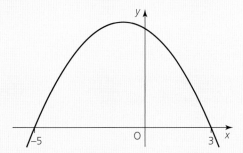

The function will be positive (due to the > symbol in the inequality) when the graph is above the x-axis.

Hence the solution is $-5 < x < 3$.

Hints & tips

If we are asked to solve the inequality $15 - 2x - x^2 < 0$ then this would occur when the graph is below the x-axis and the solution would be $x < -5$ or $x > 3$.

The intersection of a straight line and a parabola

We can find where a straight line and a parabola intersect by solving a system of equations. This will lead to a quadratic equation in which
(i) $b^2 - 4ac > 0$ (the line intersects the parabola in two distinct points),
(ii) $b^2 - 4ac = 0$ (the line is a tangent to the parabola) or
(iii) $b^2 - 4ac < 0$ (the line does not intersect the parabola).

Example

Show that the straight line with equation $y = 2x - 9$ is a tangent to the parabola with equation $y = x^2 - 4x$ and find the coordinates of the point of contact.

Solution

Substitute the linear equation into the equation of the parabola.

Hence $2x - 9 = x^2 - 4x \Rightarrow x^2 - 6x + 9 = 0$.

Calculate $b^2 - 4ac = (-6)^2 - 4 \times 1 \times 9 = 36 - 36 = 0 \Rightarrow$ the line is a tangent to the parabola.

Solve the equation $\Rightarrow x^2 - 6x + 9 = 0 \Rightarrow (x - 3)(x - 3) = 0 \Rightarrow x = 3$ (twice).

Replace x by 3 in $y = 2x - 9 \Rightarrow y = 2 \times 3 - 9 = -3$. Hence the point of contact is $(3, -3)$.

Hints & tips

Many students make a mistake when rearranging the equation at the start, so be careful. You must give an explanation why the line is a tangent, either the one given or alternatively you could say that the roots are equal after you solve the equation.

For practice

1. Express $4x^2 + 16x - 3$ in the form $a(x + p)^2 + q$.
2. What is the nature of the roots of the equation $7x^2 + 4x + 1 = 0$?
3. The graph of the function $f(x) = ax^2 - 5x + 4$ does not cross or touch the x-axis. What is the range of values of a?
4. Solve the inequality $x^2 - 5x + 6 \geq 0$.
5. Show that the line with equation $y = 4x - 5$ is a tangent to the parabola with equation $y = x^2 - 2x + 4$ and find the coordinates of the point of contact.

Chapter 5

Recurrence relations

What you should know

You should know:

★ that the formula for a linear recurrence relation is $u_{n+1} = au_n + b$
★ how to evaluate successive terms of a sequence given a recurrence relation
★ how to find a recurrence relation given three successive terms of a sequence
★ when a sequence has a limit and how to find the limit
★ how to solve problems involving recurrence relations which are in context.

Evaluating the terms in a sequence generated by a recurrence relation

A recurrence relation generates a sequence of numbers in which each term can be calculated if you know the one before. Recurrence relations can be written as $u_{n+1} = au_n + b$ where u_n represents the nth term of a sequence. When we are given the recurrence relation and one term of the sequence we can find any other term in the sequence.

Example

1 A recurrence relation is defined by $u_{n+1} = 2u_n - 3$ with $u_0 = 4$. What is the value of u_3?

Solution

$u_1 = 2u_0 - 3 = 2 \times 4 - 3 = 5$
$u_2 = 2u_1 - 3 = 2 \times 5 - 3 = 7$
$u_3 = 2u_2 - 3 = 2 \times 7 - 3 = 11$

2 A recurrence relation is defined by $u_{n+1} = 3u_n + 2$ with $u_4 = 44$. Find the value of u_2.

Solution

$u_4 = 3u_3 + 2 \Rightarrow 44 = 3u_3 + 2 \Rightarrow u_3 = \frac{44 - 2}{3} = 14$

$u_3 = 3u_2 + 2 \Rightarrow 14 = 3u_2 + 2 \Rightarrow u_2 = \frac{14 - 2}{3} = 4$

Hints & tips

Although you can evaluate the terms of a sequence very quickly using a calculator, this topic is often tested in the non-calculator paper so get used to working without one.

Finding the formula for a recurrence relation from given terms of a sequence

When you are given three terms, you can find the formula for a recurrence relation. This involves the use of simultaneous equations.

Example

A sequence is generated by the recurrence relation $u_{n+1} = au_n + b$. Find the values of a and b if $u_1 = 6$, $u_2 = 19$ and $u_3 = 58$.

Solution

$u_2 = au_1 + b \quad \Rightarrow \quad 19 = 6a + b \qquad (1)$

$u_3 = au_2 + b \quad \Rightarrow \quad 58 = 19a + b \qquad (2)$

$(2) - (1): \quad 39 = 13a \quad \Rightarrow \quad a = 3$

Substitute $a = 3$ in equation (1) leading to $\quad 19 = 6 \times 3 + b \quad \Rightarrow \quad b = 1$

Hints & tips

Remember to check your answer in this type of problem.

The limit of a sequence

In the first example, we looked at the sequence generated by the recurrence relation $u_{n+1} = 2u_n - 3$ with $u_0 = 4$. The first four terms are 4, 5, 7 and 11. Clearly the terms will get bigger and bigger as this sequence continues. This type of sequence is said to be *divergent*.

Now consider the sequence generated by the recurrence relation $u_{n+1} = 0 \cdot 5u_n + 4$ where $u_0 = 60$. Check that the first four terms are 60, 34, 21 and 14·5. In this sequence, the terms are getting closer together. Eventually (as $n \to \infty$) this sequence will reach a limit. This type of sequence is said to be *convergent*.

The sequence generated by $u_{n+1} = au_n + b$ will have a limit if $-1 < a < 1$.

We can find the limit of the sequence generated by $u_{n+1} = au_n + b$ when $-1 < a < 1$. The formula to find the limit, L, is given by $L = \dfrac{b}{1-a}$.

Example

Explain why the sequence generated by the recurrence relation $u_{n+1} = 0 \cdot 5u_n + 4$ with $u_0 = 5$ tends to a limit as $n \to \infty$. Find the limit.

Solution

The sequence has a limit because $-1 < 0 \cdot 5 < 1$.

$L = \dfrac{b}{1-a} = \dfrac{4}{1-0 \cdot 5} = \dfrac{4}{0 \cdot 5} = \dfrac{40}{5} = 8$

Hints & tips

When you are finding the limit of a sequence you should **always** justify that a limit exists as illustrated in the example above. This may be worth 1 mark. Make sure you can do this type of calculation without a calculator, notably the division by a decimal.

Problems in context

On some occasions, you will have to set up a recurrence relation before you can start to solve a problem. You may use a calculator for the following example.

Example

An industrial company has applied to the local council for permission to discharge chemical waste into a local river. It has been calculated that when the level of this chemical waste reaches 4 milligrams per litre (mg/l), then the resulting pollution level will harm fish living in the river.

A report by the local council finds that

- there is no chemical waste in the river at present
- the company wishes to discharge $2 \cdot 4$ mg/l of chemical waste into the river every week
- the natural flow of the river will remove 35% of the chemical waste every week.

The company claims that it is safe to discharge this amount of chemical waste into the river every week. Is this claim correct?

Justify your answer.

Solution

Start by setting up a recurrence relation. Each week, 35% of the chemical waste is removed, so 65% remains. Hence the new value will be $0 \cdot 65 \times$ the previous value $+ 2 \cdot 4$. This gives us the recurrence relation $u_{n+1} = 0 \cdot 65 u_n + 2 \cdot 4$.

The sequence defined by the recurrence relation has a limit because $-1 < 0 \cdot 65 < 1$.

$$L = \frac{b}{1-a} = \frac{2 \cdot 4}{1 - 0 \cdot 65} = \frac{2 \cdot 4}{0 \cdot 35} = 6 \cdot 86.$$

Because $6 \cdot 86 > 4$, the amount of chemical waste will harm the fish, therefore the claim is not correct.

For practice

1. A sequence is defined by the formula $u_{n+1} = 2u_n + 4$ with $u_0 = 3$. What is the value of u_3?

2. The sequence generated by the recurrence relation $u_{n+1} = au_n + b$ has terms $u_0 = 48$, $u_1 = 28$ and $u_2 = 18$. Find the values of a and b.

3. Find the limit of the sequence generated by the recurrence relation $u_{n+1} = 0.2u_n - 6$ with $u_0 = 3$ as $n \rightarrow \infty$.

4. A shopping centre employs a group of cleaners to remove litter daily from the premises. Every morning 90% of the litter is cleared, however each day 18 kilograms of litter is dropped.
   Comment on the long-term litter situation in the shopping centre.

Exponential and logarithmic functions

In the chapter on functions, there was a reference to the exponential and logarithmic functions.

The function $f(x) = a^x$ is called the exponential function.

The function $f(x) = \log_a x$ is called the logarithmic function.

These two functions are inverses of each other and their graphs are shown on the same diagram on page 15. In the exponential expression $N = a^x$, x is called the exponent or the logarithm of N to the base a, written $x = \log_a N$. A full relationship may then be written $N = a^x \Leftrightarrow \log_a N = x$.

This can help you get started in some questions involving logarithms. To help you remember it, you could note a particular example, i.e. $100 = 10^2 \Leftrightarrow \log_{10} 100 = 2$.

Example

1 Express $7^2 = 49$ in logarithmic form.

Solution

$\log_7 49 = 2$

2 Evaluate $\log_2\left(\dfrac{1}{8}\right)$.

Solution

Let $\log_2\left(\dfrac{1}{8}\right) = x \Rightarrow 2^x = \dfrac{1}{8} = \dfrac{1}{2^3} = 2^{-3}$, hence $x = -3$.

Graphs of exponential and logarithmic functions

Since $a^0 = 1 \Leftrightarrow \log_a 1 = 0$, we can see that, for any base a, the point $(1, 0)$ lies on the graph of $y = \log_a x$.

Since $a^1 = a \Leftrightarrow \log_a a = 1$, we can see that, for any base a, the point $(a, 1)$ lies on the graph of $y = \log_a x$.

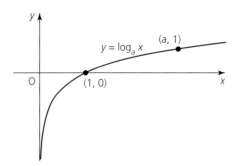

When we read $\log_a x$, we say 'the log of x to the base a'. The base, a, must be positive. The 'log' key on your calculator gives logarithms to the base 10. All logarithmic graphs lie to the right of the y-axis and pass through the points $(1, 0)$ and $(a, 1)$ where a is the base.

If $f(x) = \log_a x$, then the inverse function is $f^{-1}(x) = a^x$. As we have already seen, the graph of $f(x) = a^x$ is the image of the graph of $f(x) = \log_a x$ reflected in the line $y = x$. All exponential graphs lie above the x-axis.

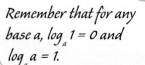
The laws of logarithms

The laws of logarithms, listed below, are true for all bases.

Key points

	Law	Illustration
1	$\log_a uv = \log_a u + \log_a v$	$\log_a 3 + \log_a 4 = \log_a(3 \times 4) = \log_a 12$
2	$\log_a\left(\dfrac{u}{v}\right) = \log_a u - \log_a v$	$\log_a 14 - \log_a 2 = \log_a\left(\dfrac{14}{2}\right) = \log_a 7$
3	$\log_a u^v = v\log_a u$	$\log_a 5^2 = 2\log_a 5$

Example

1 Express $\log_a 8 + \log_a 6 - \log_a 3$ as the logarithm of a single number.

Solution

$\log_a 8 + \log_a 6 - \log_a 3 = \log_a\left(\frac{8 \times 6}{3}\right) = \log_a\left(\frac{48}{3}\right) = \log_a 16.$ (Laws 1 and 2)

2 Evaluate $\log_2 6 + \log_2 24 - 2\log_2 3$.

Solution

$\log_2 6 + \log_2 24 - 2\log_2 3$

$= \log_2 6 + \log_2 24 - \log_2 3^2 = \log_2 6 + \log_2 24 - \log_2 9$ (Law 3)

$= \log_2\left(\frac{6 \times 24}{9}\right) = \log_2\left(\frac{144}{9}\right) = \log_2 16 = 4$ (Laws 1 and 2)

Hints & tips

An efficient way to evaluate $\log_2 16$ is to think of what power of 2 gives 16, leading to 4. In the same way, the value of $\log_2 32 = 5$ because $2^5 = 32$.

Graphs related to the basic graph of the logarithmic function

In the chapter on functions, there is a table of graphs of related functions (page 15). This will be helpful when you are asked to sketch or interpret graphs related to the basic graph of $y = \log_a x$. In addition to knowing how to sketch related graphs, you may have to use the laws of logarithms as well.

Example

Sketch the graph of $y = \log_2 16x$.

Solution

The difficulty here is deciding how to start. Use the first law of logarithms and proceed from there.

$\log_2 16x = \log_2 16 + \log_2 x = 4 + \log_2 x$ (since $16 = 2^4$).

To sketch the graph of $y = 4 + \log_2 x$, remember that $y = f(x) + a$ means that we should slide $y = f(x)$ a units up (parallel to the y-axis).

$\Rightarrow$

⇨
Hence the graph of $y = \log_2 16x$ is the same as the basic graph of $y = \log_2 x$ moved 4 units up.

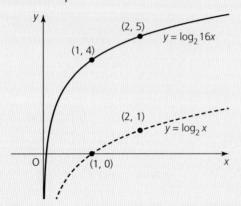

Hints & tips ⭐

Use the fact that for any base a, $\log_a 1 = 0$ and $\log_a a = 1$ to draw the graph of $y = \log_2 x$ which therefore passes through (1, 0) and (2, 1).

Example 🚩

The diagram shows the graph of $y = \log_a(x - b)$.

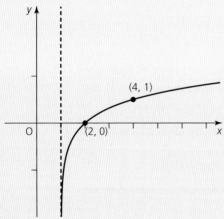

Find the values of a and b.

Solution

We know that, for any base a, the point $(1, 0)$ lies on the graph of $y = \log_a x$. As this graph passes through $(2, 0)$, we have a situation where the basic graph has been moved 1 unit to the right. From our work on related graphs, we see that the equation of the graph must be $y = \log_a(x - 1)$, hence $b = 1$. Now substitute $(4, 1)$ into this equation $\Rightarrow 1 = \log_a(4 - 1)$. Hence $1 = \log_a 3 \Rightarrow a = 3$ and the equation of the graph is $y = \log_3(x - 1)$.

You should substitute the coordinates of *both* points into your solution to check that they work.

To do well in questions on logarithmic graphs you must be knowledgeable about functions and their related graphs, the fact that for any base a, $\log_a 1 = 0$ and $\log_a a = 1$, and the laws of logarithms. It is a good idea to substitute coordinates into the equation of a graph as well.

Equations involving exponential and logarithmic expressions

Example

Solve the equation $5^x = 8$.

Solution

$$5^x = 8 \Rightarrow \log_{10} 5^x = \log_{10} 8 \Rightarrow x\log_{10} 5 = \log_{10} 8 \Rightarrow x = \frac{\log_{10} 8}{\log_{10} 5} = 1 \cdot 292$$

Hints & tips

Did you follow the working? We started by taking logarithms of both sides, then used the third law of logarithms and completed the calculation by using the 'log' key (logarithms to the base 10) on our calculator. Note that you could use the ln key that is $\frac{\ln 8}{\ln 5}$ to get the same result. We shall say more about that shortly.

Example

Solve the equation $\log_5 x = 2 \cdot 8$.

Solution

Change to exponential form and use your calculator:

$\log_5 x = 2 \cdot 8 \Rightarrow x = 5^{2 \cdot 8} = 90 \cdot 6$

Exponential decay

On your calculator, you will see the 'log' key for working out logarithms to the base 10. The inverse function is given as 10^x. You can check that these are inverse functions, for example $\log 2 = 0 \cdot 301\ldots$ and $10^{0 \cdot 301\ldots} = 2$.

Close by on your calculator you will see the 'ln' key. This works out logarithms to the base e. The inverse function is e^x. The number e is an irrational number used frequently in mathematical problems. It is approximately equal to $2 \cdot 718$. Logarithms to the base e are called natural logarithms. Check that $\ln 2 = 0 \cdot 693\ldots$ and that $e^{0 \cdot 693\ldots} = 2$. We can use natural logarithms to solve problems on *exponential decay*.

Example

The radioactive element carbon–14 can be used to estimate the age of organic remains. Carbon–14 decays according to the rule $C_t = C_0 e^{kt}$ where C_t is the amount of radioactive nuclei present at time t years and C_0 is the initial amount of radioactive nuclei.

a) The half-life of carbon–14, i.e. the time taken for half the radioactive nuclei to decay, is 5730 years. Find the value of the constant k. Give your answer correct to three significant figures.

b) What percentage of carbon–14 in a sample of bones will remain after 2000 years?

Solution

a) You can choose any value for C_0 in half-life questions, e.g. $C_0 = 100$ with $C_t = 50$. The simplest value to use is $C_0 = 1$ with $C_t = 0.5$.

Hence $C_t = C_0 e^{kt} \Rightarrow 0.5 = e^{5730k}$.

Now take natural logarithms of both sides $\Rightarrow \ln 0.5 = \ln e^{5730k}$

$\Rightarrow \ln 0.5 = 5730k \ln e \Rightarrow \ln 0.5 = 5730k$ (because $\ln e = 1$)

Hence $k = \dfrac{\ln 0.5}{5730} = -0.000121$ (correct to three sig figs).

b) The formula is now $C_t = C_0 e^{-0.000121t}$

Hence after 2000 years,

$C_t = C_0 e^{-0.000121t} = C_0 e^{(-0.000121 \times 2000)} = 0.785 C_0$.

Hence 78.5% of carbon–14 will remain in a sample of bones after 2000 years.

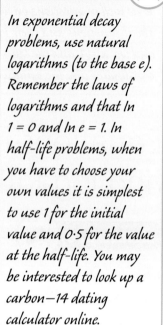

Hints & tips ⭐

In exponential decay problems, use natural logarithms (to the base e). Remember the laws of logarithms and that ln 1 = 0 and ln e = 1. In half-life problems, when you have to choose your own values it is simplest to use 1 for the initial value and 0·5 for the value at the half-life. You may be interested to look up a carbon–14 dating calculator online.

Problems on experimental data

When experiments are carried out and data is collected, it is often necessary to find a relationship or formula connecting the variables. This relationship may be linear, or perhaps exponential. In the latter case we can use logarithms to find the relationship. This leads to straight line graphs where the axes are labelled $\log x$ and $\log y$.

Example

a) Two variables x and y are connected by a relationship of the form $y = kx^n$ where k and n are constants. Prove that there is a linear relationship between $\log_{10} x$ and $\log_{10} y$.

b) In an experiment some data was obtained and a line of best fit was drawn. The table shows the data which lies on the line of best fit.

x	2	4	6	8
y	10·3	33·5	66·7	108·7

The variables x and y in the table above are connected by a relationship of the form $y = kx^n$. Find the values of k and n.

⇨

⇨
Solution

a) $y = kx^n$

Take logarithms to the base 10 of both sides, i.e. $\log_{10} y = \log_{10} kx^n$.

Hence $\log_{10} y = \log_{10} k + \log_{10} x^n = n\log_{10} x + \log_{10} k$

If we think of $X = \log_{10} x$ and $Y = \log_{10} y$, the relationship between the variables x and y can be written in the form $Y = nX + \log_{10} k$.

As this is in the form $Y = mX + c$, where $c = \log_{10} k$, there is a linear relationship between $\log_{10} x$ and $\log_{10} y$.

b) If we find the equation of the line of best fit, we will be able to find the values of k and n. We can see from the answer to part (a) that the gradient $m = n$ and the y-intercept $c = \log_{10} k$.

Consider any two values in the table, say $(2, 10·3)$ and $(8, 108·7)$.

We must take logarithms to the base 10 of the x- and y-coordinates as the line of best fit is found by plotting the logarithms of the points in the table.

Hence $(0·301, 1·013)$ and $(0·903, 2·036)$ lie on the line with equation $\log_{10} y = n\log_{10} x + \log_{10} k$. Its gradient is calculated as

$$m = \frac{y_2 - y_1}{x_2 - x_1} = \frac{2·036 - 1·013}{0·903 - 0·301} = \frac{1·023}{0·602} = 1·699.$$

Hence $n = 1·7$ (approximately)

The equation of the line is then $\log_{10} y = 1·7\log_{10} x + \log_{10} k$.

Substitute $\log_{10} x = 0·301$ and $\log_{10} y = 1·013$ into the equation.

$1·013 = 1·7 \times 0·301 + \log_{10} k \Rightarrow \log_{10} k = 1·013 - 1·7 \times 0·301 = 0·501$

As $\log_{10} k = 0·501$ then $k = 10^{0·501} = 3·17$.

Hence $k = 3·17$ and $n = 1·7$. The relationship between x and y is $y = 3·17x^{1·7}$.

> ### Hints & tips
> When two variables x and y are related by the formula $y = kx^n$, then if the graph of $\log_{10} x$ against $\log_{10} y$ is drawn it will be a straight line such that $m = n$ and $c = \log_{10} k$.

For practice

1 Evaluate $\log_2\left(\dfrac{1}{64}\right)$.

2 Simplify $\log_3 5a + \log_3 2b$.

3 Express $\log_a x^8 - \log_a x^2$ in the form $k\log_a x$.

4 Solve the equation $5^x = 130$.

5 Given that $\log_{10} x = 3\log_{10} y + \log_{10} 2$, express x in terms of y.

6 Following an experiment this graph is drawn.

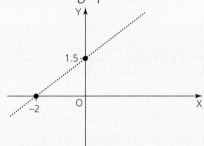

    a) Write down the equation of the line in terms of X and Y.

       It is given that $X = \log_e x$ and $Y = \log_e y$.

    b) Show that x and y satisfy a relationship of the form $y = ax^b$ and find the values of a and b.

Part 2 Geometry

Chapter 7
The straight line

What you should know

You should know:

* ★ the distance formula
* ★ how to find the midpoint of a line
* ★ how to find the gradient of a straight line
* ★ the relationship between gradient and the tangent of an angle
* ★ how to find the equation of a straight line
* ★ the relationship between perpendicular lines
* ★ the meaning of collinear points
* ★ the meaning of altitude, median and perpendicular bisector
* ★ how to find the point of intersection of two lines.

The distance formula

The distance between two points A (x_1, y_1) and B (x_2, y_2) is given by the formula $d = \sqrt{(x_2 - x_1)^2 + (y_2 - y_1)^2}$.

Example

Calculate the distance between the points $(3, -4)$ and $(-5, 2)$.

Solution

$d = \sqrt{(x_2 - x_1)^2 + (y_2 - y_1)^2} \Rightarrow d = \sqrt{(-5 - 3)^2 + (2 - (-4))^2}$

$= \sqrt{(-8)^2 + 6^2} = \sqrt{64 + 36} = \sqrt{100} = 10$

The midpoint of a straight line

To find the coordinates of the midpoint of a straight line, add the coordinates and halve the total. That is, find the mean of the x and y coordinates. Therefore the midpoint of the straight line joining the points A (x_1, y_1) and B (x_2, y_2) is given by $\left(\frac{x_1 + x_2}{2}, \frac{y_1 + y_2}{2} \right)$. Check that the midpoint of line AB in the above example is $(-1, -1)$.

The gradient formula

The gradient formula states that the gradient of the straight line joining the points (x_1, y_1) and (x_2, y_2) is given by the formula

$$m = \frac{y_2 - y_1}{x_2 - x_1}.$$

Example

Find the gradient of the straight line joining points A $(3, -4)$ and B $(-5, 2)$.

Solution

$$m_{AB} = \frac{y_2 - y_1}{x_2 - x_1} = \frac{2 - (-4)}{-5 - 3} = \frac{6}{-8} = -\frac{3}{4}$$

Hints & tips

*You should **always** simplify fractions as illustrated in the gradient example above.*
Remember that parallel lines have the same gradient.
The gradient of a line gives you information about the direction of the slope of the line (see the diagram).

Positive	Negative	Zero	Undefined
/	\	—	|

The relationship between gradient and the tangent of an angle

The gradient of a straight line may be calculated as the tangent of the angle between the straight line and the positive direction of the x-axis. We can use this fact to find the gradient from the angle and vice versa. This is often written as the formula $m = \tan\theta$ where θ is the Greek letter theta.

Example

What angle does the straight line passing through the points $(-5, 4)$ and $(3, -1)$ make with the positive direction of the x-axis?

Solution

Use the gradient formula: $m = \frac{y_2 - y_1}{x_2 - x_1} = \frac{-1 - 4}{3 - (-5)} = \frac{-5}{8} = -0 \cdot 625$, hence

$\tan\theta = -0 \cdot 625$. This leads to $\theta = 148$, so the line makes an angle of 148° with the positive direction of the x-axis. Note that the angle is obtuse because the gradient of the line is negative.

The equation of a straight line

The formula $y = mx + c$ is the equation of a straight line with gradient m passing through the point $(0, c)$. Note that c is called the y-intercept.

The formula $y - b = m(x - a)$ is the equation of a straight line with gradient m passing through the point (a, b).

Example

Find the equation of the straight line passing through the points $(-2, -3)$ and $(6, -1)$.

Solution

The gradient $= \frac{y_2 - y_1}{x_2 - x_1} = \frac{-1-(-3)}{6-(-2)} = \frac{2}{8} = \frac{1}{4}$.

The equation is $y - b = m(x - a) \Rightarrow y - (-3) = \frac{1}{4}(x - (-2)) \Rightarrow y + 3 = \frac{1}{4}(x + 2)$.

It is good form to simplify this equation by removing the fraction (multiply throughout by 4) then tidy up.

Hence $y + 3 = \frac{1}{4}(x + 2) \Rightarrow 4y + 12 = x + 2 \Rightarrow 4y = x - 10$.

Note that you could have used $(6, -1)$ for (a, b) in the formula leading to the same answer.

Hints & tips

You should be aware that the equation of the same line can appear in several different forms.
The line in the last example $4y = x - 10$ could be rearranged into the form $y = \frac{1}{4}x - \frac{5}{2}$ (can you see how this was done?). As this is in the form $y = mx + c$, we can tell that this line has gradient $\frac{1}{4}$ and crosses the y-axis at the point $\left(0, -\frac{5}{2}\right)$.
Another more general form for the equation of a straight line is given by the equation $ax + by + c = 0$. The line above would be written as $x - 4y - 10 = 0$ in this form.

Example

Find the equation of the straight line passing through the point $(3, 4)$ parallel to the line with equation $3x + 4y = 12$.

Solution

Rearrange the equation $3x + 4y = 12$ into the form $y = mx + c$ to find its gradient.

Hence $3x + 4y = 12 \Rightarrow 4y = -3x + 12 \Rightarrow y = -\frac{3}{4}x + 3 \Rightarrow m = -\frac{3}{4}$.

As parallel lines have the same gradient, the required gradient is $-\frac{3}{4}$ and we use the equation $y - b = m(x - a) \Rightarrow y - 4 = -\frac{3}{4}(x - 3) \Rightarrow 4y - 16 = -3x + 9 \Rightarrow 4y = -3x + 25$.

Hints & tips ⭐

Remember that the equation x = p where p is a constant is the equation of a line parallel to the y-axis and the equation y = q where q is a constant is the equation of a line parallel to the x-axis. Suppose you were asked to find the equation of the line joining the points (−2, 5) and (−2, −4). Check that when you use the gradient formula the denominator of the fraction is 0. As division by 0 is impossible, this indicates that the gradient is undefined and the line is vertical (parallel to the y-axis) with equation x = −2. Watch out for lines of this type.

Perpendicular lines

If two lines with non-zero gradients m_1 and m_2 are perpendicular, then $m_1 \times m_2 = -1$.

Using this rule it is simple to find the gradient of a line perpendicular to a given line. Simply turn the gradient upside down and change the sign, for example if $m_1 = \frac{2}{5}$, then a perpendicular line will have gradient $m_2 = -\frac{5}{2}$. The product of the two gradients is of course −1.

Example 🚩

PQ is a line perpendicular to the line with equation $4x - y = 8$.

P has coordinates (5, −3).

Find the equation of PQ.

Solution

Rearrange the equation $4x - y = 8$ into the form $y = 4x - 8 \Rightarrow m = 4$.

The gradient of the perpendicular line PQ is $m_{PQ} = -\frac{1}{4}$.

The equation of PQ is $y - b = m(x - a) \Rightarrow y - (-3) = -\frac{1}{4}(x - 5)$.

Use your algebra skills to simplify this equation. Well done if you got $x + 4y + 7 = 0$ or equivalent. The solution has been written in this form as it looks neater without negative signs.

Collinear points

Points which lie on the same straight line are said to be *collinear*. For example, it is clear that the points (1, 1), (2, 2) and (3, 3) are collinear. They all lie on the line with equation $y = x$. We can use gradient to show that points are collinear.

Example

Prove that the points A $(-4, -5)$, B $(0, -3)$ and C $(8, 1)$ are collinear.

Solution

$m_{AB} = \dfrac{y_2 - y_1}{x_2 - x_1} = \dfrac{-3 - (-5)}{0 - (-4)} = \dfrac{2}{4} = \dfrac{1}{2}$ and $m_{BC} = \dfrac{y_2 - y_1}{x_2 - x_1} = \dfrac{1 - (-3)}{8 - 0} = \dfrac{4}{8} = \dfrac{1}{2}$

We see that $m_{AB} = m_{BC}$. Equal gradients indicate that lines are parallel, but as point B is common to both gradients, then we can say that A, B and C are collinear.

Hints & tips

In questions such as that above (where you are asked to prove something) you should always give a complete answer. In other words, explain clearly that the lines are collinear because they have equal gradients (meaning the same direction) **and** *a common point.*

Do not confuse the word 'collinear' with 'concurrent' which is another word associated with straight lines. We say that three or more lines are concurrent if they all pass through the same point.

Some important lines

Three important lines in geometry are the *median*, the *altitude* and the *perpendicular bisector*. Make sure you do not confuse them.

The *median* is a line in a triangle from a vertex to the midpoint of the opposite side.

The *altitude* is a line in a triangle from a vertex perpendicular to the opposite side.

The *perpendicular bisector* of a line is a line passing through the midpoint and at right angles to the given line.

Example

A and B have coordinates $(4, -10)$ and $(8, 2)$ respectively. Find the equation of the perpendicular bisector of AB.

Solution

Find the midpoint of AB : $\left(\dfrac{4 + 8}{2}, \dfrac{-10 + 2}{2} \right) = (6, -4)$.

Find the gradient of AB : $m_{AB} = \dfrac{y_2 - y_1}{x_2 - x_1} = \dfrac{2 - (-10)}{8 - 4} = \dfrac{12}{4} = 3$.

Hence the gradient of the perpendicular bisector of AB $= -\dfrac{1}{3}$.

Equation of the perpendicular bisector of AB is

$y - b = m(x - a) \Rightarrow y - (-4) = -\dfrac{1}{3}(x - 6)$.

Check that this simplifies to $x + 3y + 6 = 0$.

Intersection of straight lines

If we are given the equations of two straight lines we can find the coordinates of the point where the lines intersect by solving simultaneous equations. It is important that you can solve simultaneous equations accurately and consistently. For practice, try to find the point of intersection of the lines with equations $x + 3y = 9$ and $2x - y = 4$ using simultaneous equations. The correct answer will be given later.

Now that we have studied a large variety of facts regarding the straight line we can attempt longer and more complicated questions. Before tackling them, it should be pointed out that many students make avoidable errors, often right at the start of such questions and this means that marks are lost unnecessarily.

Hints & tips

Always check that you have copied down the correct figures and equations at the start of a question. In long questions worth nine or ten marks, copying wrong figures or equations will lead quickly to solutions which don't work out properly. At best this will take time to spot and put right, at worst it will cause distress and lead to negative feelings. Other common errors occur when rearranging an equation into the form y = mx + c, when subtracting negative numbers, and when making basic mistakes with formulae such as the distance formula, the gradient formula and the midpoint formula. The best way to eliminate such errors is plenty of practice.

(The answer to the simultaneous equations question above was (3, 2). Well done if you got that!)

Next we shall look at a longer, more complicated example. This would probably be worth nine marks in an examination (three for each part), so check every step carefully and try to eliminate avoidable errors when you tackle similar problems on your own.

Example

Triangle DEF has vertices D (−2, 2), E (8, 4) and F (2, −5).
   a) Find the equation of the median FM.
   b) Find the equation of the altitude DN.
   c) Find the coordinates of G, the point of intersection of FM and DN.

Solution

If a diagram is not provided, it is a good idea to make a rough sketch on plain paper. You can use your sketch to check if your answers seem sensible and if the signs of the gradients are correct.

⇨

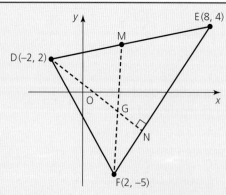

a) The median FM joins F to M, the midpoint of DE.

As D is $(-2, 2)$ and E is $(8, 4)$ M is $\left(\dfrac{-2+8}{2}, \dfrac{2+4}{2}\right) = (3, 3)$.

The gradient of FM $= \dfrac{y_2 - y_1}{x_2 - x_1} = \dfrac{3 - (-5)}{3 - 2} = \dfrac{8}{1} = 8$.

The equation of FM is $y - b = m(x - a) \Rightarrow y + 5 = 8(x - 2)$.

This simplifies to $y + 5 = 8x - 16 \Rightarrow 8x - y = 21$.

b) The altitude DN is perpendicular to EF.

The gradient of EF $= \dfrac{-5 - 4}{2 - 8} = \dfrac{-9}{-6} = \dfrac{3}{2}$.

Therefore the gradient of DN is $-\dfrac{2}{3}$.

The equation of DN is $y - b = m(x - a) \Rightarrow y - 2 = -\dfrac{2}{3}(x + 2)$.

This simplifies to $3y - 6 = -2x - 4 \Rightarrow 2x + 3y = 2$.

c) To find the point of intersection of FM and DN, solve simultaneous equations.

$$8x - y = 21 \qquad (1)$$
$$2x + 3y = 2 \qquad (2)$$

$(1) \times 3$: $\qquad\qquad 24x - 3y = 63 \qquad (3)$

$(2) + (3)$: $\qquad\qquad 26x = 65 \Rightarrow x = 2\cdot5$

By substitution in (1): $8 \times 2\cdot5 - y = 21 \Rightarrow y = -1$

Hence the coordinates of G are $(2\cdot5, -1)$.

For practice

1 Calculate the distance between the points A $(-8, 6)$ and B $(4, 1)$.

2 Find the equation of the straight line passing through the point $(3, -4)$ which is parallel to the line with equation $3y - 6x = 5$.

3 Find the equation of the straight line perpendicular to the line $2y = 3x + 4$ and passing through the point $(8, 2)$.

4 A straight line has gradient $-\sqrt{3}$. What angle does this line make with the positive direction of the x-axis?

5 Find the equation of the perpendicular bisector of the line joining the points $(4, 6)$ and $(10, -2)$.

6 P, Q and R are the points $(4, 7)$, $(-3, 0)$ and $(5, -2)$ respectively. M is the midpoint of PR.

   a) Calculate the coordinates of M.

   b) Find the equation of QM.

   c) The perpendicular from R to PQ meets QM in T. Find the coordinates of T.

The circle

What you should know

You should know:

★ that the circle, centre O, radius r, has equation $x^2 + y^2 = r^2$
★ that the circle, centre (a, b), radius r, has equation $(x - a)^2 + (y - b)^2 = r^2$
★ that the equation $x^2 + y^2 + 2gx + 2fy + c = 0$ is the equation of a circle with centre $(-g, -f)$ and radius $\sqrt{g^2 + f^2 - c}$
★ how to find the equation of a tangent to a circle at a given point on the circumference
★ how to find the points of intersection of a straight line and a circle
★ how to verify whether a straight line is a tangent to a circle or not.

The circle with centre O and radius r

The circle, centre O, radius r, has equation $x^2 + y^2 = r^2$.

As an example, the circle with equation $x^2 + y^2 = 16$ has centre O and radius 4. Similarly, a circle with centre O and radius 9 has equation $x^2 + y^2 = 81$.

The circle with centre (a, b) and radius r

The circle, centre (a, b), radius r, has equation $(x - a)^2 + (y - b)^2 = r^2$.

As an example, the circle with equation $(x - 3)^2 + (y + 2)^2 = 25$ has centre $(3, -2)$ and radius 5.

Similarly, a circle with centre $(-4, 0)$ and radius 7 has equation $(x + 4)^2 + y^2 = 49$.

Example

A circle has the line joining A $(-4, 3)$ to B $(6, -5)$ as its diameter.
What is the equation of the circle?

Solution

As AB is a diameter of the circle, the centre is the midpoint of AB. (Check that the centre is at $(1, -1)$.) The radius can be found by calculating the distance from the centre to A (or B) using the distance formula. We shall find the distance from $(1, -1)$ to A $(-4, 3)$.

Radius

$$r = \sqrt{(x_2 - x_1)^2 + (y_2 - y_1)^2} = \sqrt{(-4-1)^2 + (3+1)^2} = \sqrt{25 + 16} = \sqrt{41}$$

Now we know the centre and radius, use the formula
$(x - a)^2 + (y - b)^2 = r^2$.

Hence the equation of the circle is $(x - 1)^2 + (y + 1)^2 = 41$.

Hints & tips

If you are asked to find the equation of a circle, use the formula $(x - a)^2 + (y - b)^2 = r^2$. You may have to use the distance formula to find the radius.

The equation $x^2 + y^2 + 2gx + 2fy + c = 0$

The equation $x^2 + y^2 + 2gx + 2fy + c = 0$ is the equation of a circle with centre $(-g, -f)$ and radius $\sqrt{g^2 + f^2 - c}$ provided that $g^2 + f^2 - c > 0$.

This equation is called *the general equation of a circle.*

Example

Find the centre and radius of the circle with equation
$x^2 + y^2 + 12x - 6y + 36 = 0$.

Solution

$2g = 12 \Rightarrow g = 6 \Rightarrow -g = -6$ and $2f = -6 \Rightarrow f = -3 \Rightarrow -f = 3$
so the centre is $(-6, 3)$.

Radius $r = \sqrt{g^2 + f^2 - c} = \sqrt{6^2 + (-3)^2 - 36} = \sqrt{36 + 9 - 36} = \sqrt{9} = 3.$

Hints & tips

Normally it should be a simple matter to write down the centre without any working. Many extended questions on the circle start with an example like the one above. So be careful as an error at the start would cause big problems later on.

Example

Two circles have equations $x^2 + y^2 + 4x + 8y + 15 = 0$ and
$x^2 + y^2 - 12x - 9 = 0$. Show that these circles touch externally.

Solution

The circle with equation $x^2 + y^2 + 4x + 8y + 15 = 0$ has centre $(-2, -4)$.

Its radius $= \sqrt{g^2 + f^2 - c} = \sqrt{2^2 + 4^2 - 15} = \sqrt{4 + 16 - 15} = \sqrt{5}$.

The circle with equation $x^2 + y^2 - 12x - 9 = 0$ has centre $(6, 0)$.

Its radius

$= \sqrt{g^2 + f^2 - c} = \sqrt{(-6)^2 + 0^2 - (-9)} = \sqrt{36 + 9} = \sqrt{45} = \sqrt{9 \times 5} = 3\sqrt{5}$.

Use the distance formula to find the distance between the centres $(-2, -4)$ and $(6, 0)$.

$d = \sqrt{(x_2 - x_1)^2 + (y_2 - y_1)^2} = \sqrt{(6 + 2)^2 + (0 + 4)^2} = \sqrt{64 + 16}$

$= \sqrt{80} = \sqrt{16 \times 5} = 4\sqrt{5}$

Since the sum of the radii $\left(\sqrt{5} + 3\sqrt{5} = 4\sqrt{5}\right)$ equals the distance between the centres, the circles must touch externally.

Hints & tips

It is possible to work out the coordinates of the point where the two circles touch. As the radii are in the ratio 1 : 3, this can be done using the section formula discussed later in vectors. The two circles touch at (0, −3). Try to verify this.

Finding the equation of a tangent to a circle

Example

a) Show that the point $(7, 2)$ lies on the circle
$x^2 + y^2 - 6x + 4y - 19 = 0$.

b) Find the equation of the tangent at the point $(7, 2)$ on this circle.

Solution

a) Substitute $x = 7$, $y = 2$ into $x^2 + y^2 - 6x + 4y - 19 = 0$.
Hence $7^2 + 2^2 - 6 \times 7 + 4 \times 2 - 19 = 49 + 4 - 42 + 8 - 19 = 0$.
Therefore the point $(7, 2)$ lies on the circle.

b) Centre of the circle is $(3, -2)$ (by inspection).

Find the gradient of the radius: $m_{radius} = \dfrac{y_2 - y_1}{x_2 - x_1} = \dfrac{-2 - 2}{3 - 7} = \dfrac{-4}{-4} = 1$.

As the tangent is perpendicular to the radius $m_{tangent} = -1$.
Use the formula $y - b = m(x - a)$ to find equation of tangent.
Hence equation is $y - 2 = -1(x - 7) \Rightarrow y - 2 = -x + 7 \Rightarrow x + y = 9$.

Finding the points of intersection of a straight line and a circle

We can find where a straight line and a circle intersect by solving a system of equations in the same way as we did with a straight line and a parabola.

Example

Find the points of intersection of the line with equation $5x = 20 - y$ and the circle with equation $x^2 + y^2 - 2x - 4y - 8 = 0$.

Solution

Re-arrange the equation of the straight line to make x or y the subject of the equation. Choose y this time in order to avoid fractions in your calculation. Be careful with signs as an error at this stage means that you will give yourself a lot of extra work.

$5x = 20 - y \Rightarrow y = 20 - 5x$

Substitute $y = 20 - 5x$ into $x^2 + y^2 - 2x - 4y - 8 = 0$.

$x^2 + (20 - 5x)^2 - 2x - 4(20 - 5x) - 8 = 0$

$\Rightarrow x^2 + 400 - 200x + 25x^2 - 2x - 80 + 20x - 8 = 0$

$\Rightarrow 26x^2 - 182x + 312 = 0$ (this looks difficult so try to find a common factor)

$\Rightarrow x^2 - 7x + 12 = 0$ (by dividing throughout by 26)

$\Rightarrow (x - 3)(x - 4) = 0 \Rightarrow x = 3$ or $x = 4$.

Substitute $x = 3$ and $x = 4$ into $y = 20 - 5x$ to find points of intersection.

Hence the points of intersection are $(3, 5)$ and $(4, 0)$.

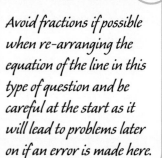

Hints & tips

Avoid fractions if possible when re-arranging the equation of the line in this type of question and be careful at the start as it will lead to problems later on if an error is made here.

Circles and tangents

When we find where a straight line and a circle intersect, we solve a system of equations. This will lead to a quadratic equation.

If $b^2 - 4ac > 0$, then the line intersects the circle in two distinct points.

If $b^2 - 4ac = 0$, the line is a tangent to the circle.

If $b^2 - 4ac < 0$, the line does not intersect the circle.

We shall look at the case where a line is a tangent to a circle. In this case, when you arrive at the quadratic equation, not only will $b^2 - 4ac = 0$, but you will get two equal solutions when you solve the equation. If you are asked to prove that a line is a tangent to a circle, you *must* give a reason.

Example

Show that the straight line with equation $2x + y = 9$ is a tangent to the circle with equation $x^2 + y^2 - 2x - 4y = 0$.

Solution

$2x + y = 9 \Rightarrow y = 9 - 2x$

Substitute $y = 9 - 2x$ into $x^2 + y^2 - 2x - 4y = 0$

Hence $x^2 + (9 - 2x)^2 - 2x - 4(9 - 2x) = 0$

$\Rightarrow x^2 + 81 - 36x + 4x^2 - 2x - 36 + 8x = 0$

$\Rightarrow 5x^2 - 30x + 45 = 0$ (divide throughout by 5)

$\Rightarrow x^2 - 6x + 9 = 0$

$\Rightarrow b^2 - 4ac = (-6)^2 - 4 \times 1 \times 9 = 36 - 36 = 0$

$\Rightarrow$ the straight line is a tangent to the circle.

If you are asked to find the point of contact, you can solve the equation.

$x^2 - 6x + 9 = 0 \Rightarrow (x - 3)(x - 3) = 0 \Rightarrow x = 3$

When $x = 3$, $y = 9 - 2 \times 3 = 3$, hence the point of contact is (3, 3).

The fact that the roots are equal confirms that the straight line is a tangent to the circle.

Hints & tips

The reason you give for a line being a tangent to a circle can either be that $b^2 - 4ac = 0$ or that the equation has equal roots.

For practice

1 A circle has equation $2x^2 + 2y^2 + 12x - 8y - 46 = 0$. Find the centre and radius of this circle.

2 A circle with centre (4, −1) passes through the point (1, 3). What is the equation of this circle?

3 Find the equation of the tangent to the circle with equation $x^2 + y^2 = 25$ at the point (−3, 4) on its circumference.

4 A circle has equation $x^2 + y^2 - 6x - 10y + 26 = 0$.
   a) State the coordinates of C, the centre of the circle.
   b) Find the equation of the chord with midpoint M (4, 6).
   c) Show that the line with equation $x + y = 12$ is a tangent to the circle and find the coordinates of P, the point of contact.

5 A circle passes through the origin, P (0, 6) and Q (8, 0). Find its equation.
   HINT: The angle in a semi-circle is a right angle.

45

Vectors

You should know:
- ★ the meaning of the terms components and position vector
- ★ how to use vector pathways
- ★ how to calculate the magnitude of a vector
- ★ how to use the distance formula in 3-dimensions
- ★ the meaning of a unit vector
- ★ how to find whether three points in space are collinear
- ★ how to use the section formula
- ★ that the scalar product is given by $\mathbf{a.b} = |\mathbf{a}||\mathbf{b}|\cos\theta$, where θ is the angle between $\mathbf{a}$ and $\mathbf{b}$
- ★ how to calculate the angle between two vectors
- ★ how to use the distributive law with vectors.

Position vectors

A position vector shows how to arrive at a point from the origin O. So if a point in space, A, has coordinates (x, y, z) then point A can be represented by the position vector $\mathbf{a} = \begin{pmatrix} x \\ y \\ z \end{pmatrix}$ where $\mathbf{a} = \overrightarrow{OA}$.

The components of a vector are written in a column and must not be confused with coordinates. Coordinates tell us the position of a point, while components show a shift in a particular direction. If we know the coordinates of two points, A and B, we can find the components of the vector $\overrightarrow{AB}$ using the rule $\overrightarrow{AB} = \mathbf{b} - \mathbf{a}$.

Example

A and B have coordinates $(2, -3, -5)$ and $(-1, 1, 7)$ respectively. Find the components of the vector $\overrightarrow{AB}$.

Solution

$$\overrightarrow{AB} = \mathbf{b} - \mathbf{a} = \begin{pmatrix} -1 \\ 1 \\ 7 \end{pmatrix} - \begin{pmatrix} 2 \\ -3 \\ -5 \end{pmatrix} = \begin{pmatrix} -3 \\ 4 \\ 12 \end{pmatrix}$$

The magnitude of a vector

The magnitude (or length) of a vector $\mathbf{u}$ is denoted by $|\mathbf{u}|$. If $\mathbf{u} = \begin{pmatrix} x \\ y \\ z \end{pmatrix}$, then $|\mathbf{u}| = \sqrt{x^2 + y^2 + z^2}$.

Example

Find the magnitude of vector $\overrightarrow{AB}$ from the previous example.

Solution

$$\overrightarrow{AB} = \begin{pmatrix} -3 \\ 4 \\ 12 \end{pmatrix} \Rightarrow |\overrightarrow{AB}| = \sqrt{(-3)^2 + 4^2 + 12^2} = \sqrt{9 + 16 + 144} = \sqrt{169} = 13$$

Hints & tips

*In textbooks and examination papers, vectors appear as directed line segments, for example $\overrightarrow{AB}$, or in bold, for example **u**. As you cannot write in bold on an examination script or in a jotter, you should always underline vectors, for example you should write **u** as u.*

Note that it would not make sense to write $\vec{A}$ as directed line segments must have two letters.

The magnitude of vector $\overrightarrow{AB}$ can be written as $|\overrightarrow{AB}|$ or simply AB, but never as $\overrightarrow{AB}$.

*The magnitude of vector **u** should be written as $|u|$ but never as **u**.*

The distance formula

The distance between two points $A (x_1, y_1, z_1)$ and $B (x_2, y_2, z_2)$ is

given by the formula $d = \sqrt{(x_2 - x_1)^2 + (y_2 - y_1)^2 + (z_2 - z_1)^2}$.

Example

Calculate the distance between the points (3, −4, 1) and (−5, 2, 6). Express your answer as a surd in its simplest form.

Solution

$$d = \sqrt{(x_2 - x_1)^2 + (y_2 - y_1)^2 + (z_2 - z_1)^2}$$

$$\Rightarrow d = \sqrt{(-5 - 3)^2 + (2 + 4)^2 + (6 - 1)^2}$$

$$= \sqrt{(-8)^2 + 6^2 + 5^2} = \sqrt{64 + 36 + 25} = \sqrt{125} = \sqrt{25 \times 5} = 5\sqrt{5}.$$

Vector pathways

Pathways in vectors can be used to solve many problems.

> ### Example
>
> PQRS is a parallelogram. T is a point on RS such that RT : TS = 1 : 2.
>
>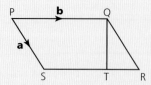
>
> Given that $\overrightarrow{PS}$ and $\overrightarrow{PQ}$ represent the vectors **a** and **b** respectively, express $\overrightarrow{QT}$ in terms of **a** and **b**.
>
> ### Solution
>
> If T is a point on RS such that RT : TS = 1 : 2, then $\overrightarrow{RT} = \frac{1}{3}\overrightarrow{RS}$.
>
> $\overrightarrow{QT} = \overrightarrow{QR} + \overrightarrow{RT} = \overrightarrow{QR} + \frac{1}{3}\overrightarrow{RS} = \mathbf{a} - \frac{1}{3}\mathbf{b}$.

> **Hints & tips** ⭐
>
> *When looking at pathways in a vector diagram, remember that vectors in the same direction as the arrow are positive while vectors in the opposite direction to the arrow are negative.*

Unit vectors

A *unit vector* is any vector which has length 1 unit.

Three important unit vectors are **i**, **j** and **k**. These are three vectors, each of length 1 unit, parallel to the *x*-, *y*- and *z*-axes respectively.

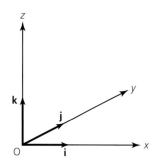

Hence $\mathbf{i} = \begin{pmatrix} 1 \\ 0 \\ 0 \end{pmatrix}$, $\mathbf{j} = \begin{pmatrix} 0 \\ 1 \\ 0 \end{pmatrix}$ and $\mathbf{k} = \begin{pmatrix} 0 \\ 0 \\ 1 \end{pmatrix}$. If a point P has coordinates (a, b, c),

then the vector $\overrightarrow{OP} = \mathbf{p} = a\mathbf{i} + b\mathbf{j} + c\mathbf{k}$ and has components $\begin{pmatrix} a \\ b \\ c \end{pmatrix}$.

> ### Example
>
> Find the length of vector $\mathbf{v} = 3\mathbf{i} - 2\mathbf{j} + 4\mathbf{k}$.
>
> ### Solution
>
> The vector $3\mathbf{i} - 2\mathbf{j} + 4\mathbf{k}$ has components $\begin{pmatrix} 3 \\ -2 \\ 4 \end{pmatrix}$.
>
> Hence $|\mathbf{v}| = \sqrt{3^2 + (-2)^2 + 4^2} = \sqrt{9 + 4 + 16} = \sqrt{29}$.

Example

If $\frac{1}{2}\mathbf{i} + \frac{1}{4}\mathbf{j} + a\mathbf{k}$ is a unit vector, find the values of a.

Solution

Here $\left(\frac{1}{2}\right)^2 + \left(\frac{1}{4}\right)^2 + a^2 = 1 \Rightarrow \frac{1}{4} + \frac{1}{16} + a^2 = 1 \Rightarrow a^2 = 1 - \frac{5}{16} = \frac{11}{16} \Rightarrow a = \pm\sqrt{\frac{11}{16}}$.

Collinear points

In general, vectors $\mathbf{u}$ and $k\mathbf{u}$, where k is a constant, are vectors in the same direction such that the magnitude of vector $k\mathbf{u}$ is $k \times$ the magnitude of vector $\mathbf{u}$. Vectors $\mathbf{u}$ and $k\mathbf{u}$ will be parallel. If they share a common point, they will be collinear.

We looked earlier at collinear points in the chapter on the straight line. Then we proved that three points were collinear using gradients. For points in space, we use vectors.

Example

Show that the points A $(2, -1, 6)$, B $(6, 1, 2)$ and C $(14, 5, -6)$ are collinear and find the ratio in which B divides AC.

Solution

$$\overrightarrow{AB} = \mathbf{b} - \mathbf{a} = \begin{pmatrix} 6 \\ 1 \\ 2 \end{pmatrix} - \begin{pmatrix} 2 \\ -1 \\ 6 \end{pmatrix} = \begin{pmatrix} 4 \\ 2 \\ -4 \end{pmatrix} \text{ and } \overrightarrow{BC} = \mathbf{c} - \mathbf{b} = \begin{pmatrix} 14 \\ 5 \\ -6 \end{pmatrix} - \begin{pmatrix} 6 \\ 1 \\ 2 \end{pmatrix} = \begin{pmatrix} 8 \\ 4 \\ -8 \end{pmatrix}$$

$\overrightarrow{AB} = \frac{1}{2}\overrightarrow{BC}$ so AB is parallel to BC with point B in common. Hence A, B and C are collinear. The ratio in which B divides AC is $1 : 2$.

Note that it is easy to check you have the correct ratio by looking at the x-coordinates of A, B and C. Check that they go $2 \rightarrow 6 \rightarrow 14$. This shows a jump of 4 followed by a jump of 8 leading to the ratio $4 : 8 = 1 : 2$.

Hints & tips

*If you are asked to show that three points in space are collinear in your examination, remember to communicate properly by mentioning parallel **and** common point.*

The section formula

The section formula states that if a point P divides a line AB in the ratio $m : n$, then

$$\mathbf{p} = \frac{m\mathbf{b} + n\mathbf{a}}{m + n}.$$

Note that this formula does not appear on the list of formulae in your examination. A particular case of this formula occurs when a point M is the midpoint of AB (and therefore divides AB in the ratio $1 : 1$). Can you see that $\mathbf{m} = \frac{1}{2}(\mathbf{a} + \mathbf{b})$? This result is useful to remember.

Example

A and B are the points (4, 4, 10) and (10, −2, 4) respectively.

The point P lies between A and B and divides AB in the ratio 1 : 2.

Find the coordinates of P.

Solution

$$\mathbf{p} = \frac{m\mathbf{b} + n\mathbf{a}}{m+n} = \frac{1}{1+2}\left[1\begin{pmatrix} 10 \\ -2 \\ 4 \end{pmatrix} + 2\begin{pmatrix} 4 \\ 4 \\ 10 \end{pmatrix}\right] = \frac{1}{3}\left[\begin{pmatrix} 10 \\ -2 \\ 4 \end{pmatrix} + \begin{pmatrix} 8 \\ 8 \\ 20 \end{pmatrix}\right] = \frac{1}{3}\begin{pmatrix} 18 \\ 6 \\ 24 \end{pmatrix} = \begin{pmatrix} 6 \\ 2 \\ 8 \end{pmatrix}$$

Hence the coordinates of P = (6, 2, 8).

In practice, it is probably simpler to use a sketch with crossing lines for the section formula. Check the following working carefully.

A (4, 4, 10)   B (10, −2, 4)

1 : 2

P is $\left(\dfrac{2 \times 4 + 1 \times 10}{3}, \dfrac{2 \times 4 + 1 \times (-2)}{3}, \dfrac{2 \times 10 + 1 \times 4}{3}\right) = (6, 2, 8)$.

The scalar product

The scalar product (or dot product) of two vectors is given by the formula $\mathbf{a}.\mathbf{b} = |\mathbf{a}||\mathbf{b}|\cos\theta$, where θ is the angle between vectors $\mathbf{a}$ and $\mathbf{b}$. As the name suggests, the scalar product of two vectors is not itself a vector, but a scalar, i.e. a number.

Example

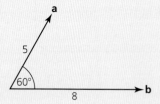

Evaluate **a.b**

Solution

Here $\mathbf{a}.\mathbf{b} = |\mathbf{a}||\mathbf{b}|\cos\theta = 5 \times 8 \times \cos 60° = 5 \times 8 \times 0 \cdot 5 = 20$.

Hints & tips

If you have to use the formula above, check the diagram to ensure that both arrows are pointing away from the vertex of the angle or both arrows are pointing towards the vertex. Otherwise your solution will be the negative of the correct solution.

The scalar product can also be given in component form using the formula

$$\mathbf{a.b} = a_1b_1 + a_2b_2 + a_3b_3 \text{ where } \mathbf{a} = \begin{pmatrix} a_1 \\ a_2 \\ a_3 \end{pmatrix} \text{ and } \mathbf{b} = \begin{pmatrix} b_1 \\ b_2 \\ b_3 \end{pmatrix}.$$

Both versions of the formula for the scalar product are given in the list of formulae.

Example

Find the value of **u.v** where **u** = 3**i** − 2**j** + 4**k** and **v** = 4**i** − 3**j**.

Solution

$$\mathbf{u} = 3\mathbf{i} - 2\mathbf{j} + 4\mathbf{k} = \begin{pmatrix} 3 \\ -2 \\ 4 \end{pmatrix} \text{ and } \mathbf{v} = 4\mathbf{i} - 3\mathbf{j} = \begin{pmatrix} 4 \\ -3 \\ 0 \end{pmatrix}$$

$$\mathbf{u.v} = (3 \times 4) + \left[(-2) \times (-3)\right] + (4 \times 0) = 12 + 6 + 0 = 18$$

Calculating the angle between two vectors

We calculate the angle between two vectors by first changing the subject of the formula in $\mathbf{a.b} = |\mathbf{a}||\mathbf{b}|\cos\theta$ to $\cos\theta$ leading to $\cos\theta = \dfrac{\mathbf{a.b}}{|\mathbf{a}||\mathbf{b}|}$.

Example

Calculate the size of the angle between **u** and **v** where **u** = 3**i** − 2**j** + 4**k** and **v** = 4**i** − 3**j**.

Solution

$$\mathbf{u.v} = (3 \times 4) + \left[(-2) \times (-3)\right] + (4 \times 0) = 12 + 6 + 0 = 18 \text{ (as above)}$$

$$|\mathbf{u}| = \sqrt{3^2 + (-2)^2 + 4^2} = \sqrt{9 + 4 + 16} = \sqrt{29}$$

$$|\mathbf{v}| = \sqrt{4^2 + (-3)^2} = \sqrt{16 + 9} = \sqrt{25} = 5$$

Let the angle between **u** and **v** be θ.

$$\cos\theta = \frac{\mathbf{u.v}}{|\mathbf{u}||\mathbf{v}|} = \frac{18}{\sqrt{29} \times 5} = 0 \cdot 669 \Rightarrow \theta = 48$$

Hence the angle between **u** and **v** is 48°.

Using the distributive law with vectors

The scalar product is distributive over addition, that is $\mathbf{a}.(\mathbf{b} + \mathbf{c}) = \mathbf{a}.\mathbf{b} + \mathbf{a}.\mathbf{c}$.

Example

Vectors $\mathbf{u}$, $\mathbf{v}$ and $\mathbf{w}$ are shown in the diagram and angle DCB = 30°.

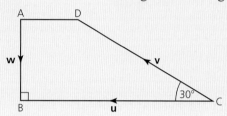

Given that $|\mathbf{u}| = 12$ and $|\mathbf{v}| = 8$, find the exact value of $\mathbf{u}.(\mathbf{v} + \mathbf{w})$.

Solution

$\mathbf{u}.(\mathbf{v} + \mathbf{w}) = \mathbf{u}.\mathbf{v} + \mathbf{u}.\mathbf{w} =$

$|\mathbf{u}||\mathbf{v}|\cos 30° + |\mathbf{u}||\mathbf{w}|\cos 90° = 12 \times 8 \times \dfrac{\sqrt{3}}{2} + 0 = 48\sqrt{3}$

Hints & tips

If you calculate the value of the cosine to be negative, the angle is obtuse.

Example

For what value of k are the vectors $\mathbf{a} = \begin{pmatrix} 3 \\ -2 \\ k \end{pmatrix}$ and $\mathbf{b} = \begin{pmatrix} 4 \\ -5 \\ 2 \end{pmatrix}$ perpendicular?

Solution

$\mathbf{a}$ and $\mathbf{b}$ are perpendicular if $\mathbf{a}.\mathbf{b} = 0$

Hence

$(3 \times 4) + [(-2) \times (-5)] + (2 \times k) = 0 \Rightarrow 12 + 10 + 2k = 0 \Rightarrow k = -11.$

Hints & tips

Note that $\mathbf{a}.\mathbf{a} = |a||a|\cos 0° = |a|^2$ for any vector a as $\cos 0° = 1$. When two vectors a and b are perpendicular $\mathbf{a}.\mathbf{b} = 0$ because $\cos 90° = 0$.

For practice

1 Calculate the distance between the points $(-2, 4, 0)$ and $(3, -1, 2)$.
2 Find the length of vector $\mathbf{u} = \mathbf{i} - 4\mathbf{j} + 2\mathbf{k}$.
3 Show that the points A $(1, -2, 9)$, B $(5, 0, 5)$ and C $(17, 6, -7)$ are collinear, and find the ratio in which B divides AC.
4 P and Q are the points $(3, -4, 2)$ and $(15, 12, -10)$ respectively. The point T lies between P and Q and divides PQ in the ratio 3:1. Find the coordinates of T.
5 A triangle ABC has vertices A $(3, -2, 1)$, B $(0, 5, -4)$ and C $(3, 6, -1)$.
   a) Find the components of $\overrightarrow{AB}$.
   b) Find the components of $\overrightarrow{AC}$.
   c) Calculate the size of angle BAC.

Part 3 Trigonometry

Chapter 10
Basic trigonometry

What you should know

You should know:
★ the formulae $\sin^2 x + \cos^2 x = 1$ and $\tan x = \dfrac{\sin x}{\cos x}$
★ the exact values of sin, cos and tan of 30°, 45° and 60°
★ the basic trigonometric graphs, their amplitudes and periods
★ the meaning of phase angle
★ how to convert from degrees to radians and vice versa
★ how to solve basic trigonometric equations
★ how to prove trigonometric identities.

Two important formulae

You should be familiar with two important formulae which are useful in proving trigonometric identities.

$\sin^2 x + \cos^2 x = 1$ and $\tan x = \dfrac{\sin x}{\cos x}$

We can rearrange the first formula into $\sin^2 x = 1 - \cos^2 x$ or $\cos^2 x = 1 - \sin^2 x$.

Exact values

You must learn the exact values of sin, cos and tan of 30°, 45° and 60°. These values are often tested in the non-calculator paper in your examination. You should also be very familiar with 'all, sin, tan, cos' in the quadrants and the related angles in the quadrants.

	30°	45°	60°
sin	$\dfrac{1}{2}$	$\dfrac{1}{\sqrt{2}}$	$\dfrac{\sqrt{3}}{2}$
cos	$\dfrac{\sqrt{3}}{2}$	$\dfrac{1}{\sqrt{2}}$	$\dfrac{1}{2}$
tan	$\dfrac{1}{\sqrt{3}}$	1	$\sqrt{3}$

Example

1 What is the exact value of cos 210°?

Solution

$$\cos 210° = -\cos 30° = -\frac{\sqrt{3}}{2}$$

2 Solve the equation $\tan x° = \sqrt{3}$, $0 \leq x \leq 360$.

Solution

As the tangent ratio is positive in the 1st and 3rd quadrants, $x = 60$ or 240.

Hints & tips

Remember that examples of the type above are non-calculator, so check that you have memorised the values in the table or can deduce them from diagrams.

Basic trigonometric graphs

Knowledge of the basic sine and cosine graphs will prove essential.

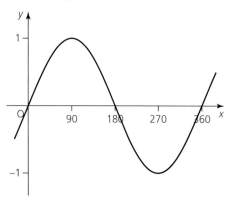

The sine graph

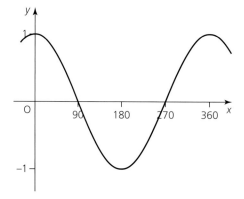

The cosine graph

By considering the graphs, we can quickly identify key values involving multiples of 90°. Check that $\sin 0° = 0$, $\sin 90° = 1$, $\sin 180° = 0$, $\sin 270° = -1$, $\sin 360° = 0$ and $\cos 0° = 1$, $\cos 90° = 0$, $\cos 180° = -1$, $\cos 270° = 0$ and $\cos 360° = 1$.

You should also be familiar with the tangent graph shown.

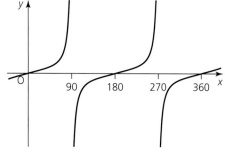

The tangent graph

Example

Sketch the graph of $y = 2\sin(x - 45)°$, $-360 \leq x \leq 360$.

Solution

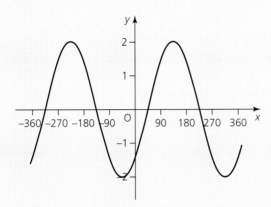

We can see that the basic sine graph $y = \sin x°$ moved 45° to the right and stretched parallel to the y-axis by a scale factor of 2. The *amplitude* of the graph (half the difference between the maximum and minimum values) is 2 and the *phase angle* is 45°. The graph has a repeating pattern. The period of the graph is length of one cycle of the graph. Check that the period of $y = 2\sin(x - 45)°$ is 360°.

Radian measure

Angles can be measured in degrees or radians. You should be able to convert from one to the other. The radian is a larger unit of angular measurement than the degree. One radian is the angle subtended at the centre of a circle by an arc equal in length to the radius. One radian is equal to about 57°.

This is done using direct proportion and the fact that π radians = 180°.

Example

1 Convert 225° to radians.

Solution

$225° = \dfrac{225}{180}\pi = \dfrac{5\pi}{4}$ radians   (always simplify the fraction)

2 Convert $\dfrac{11\pi}{6}$ radians to degrees.

Solution

$\dfrac{11\pi}{6} = \dfrac{11}{6} \times 180° = 330°$

Hints & tips

If you remember that π radians = 180°, then you will be able to convert from degrees to radians and vice versa.

Trigonometric equations

You should be familiar with solving simple trigonometric equations of the type $2 \sin x° - 1 = 0, 0 \leq x \leq 360$. Check that the solutions to this equation are $x = 30$ or 150. However, it is now expected that you should be able to solve more difficult equations such as $2 \cos^2 x° + \cos x° - 1 = 0, 0 \leq x \leq 360$ and $2 \cos 3x° = 1, 0 \leq x \leq 360$. The first of these two equations is a quadratic equation in $\cos x°$ and can be solved by factorisation, while the second involves a multiple angle.

Example

Solve the equation $2 \cos^2 x° + \cos x° - 1 = 0, 0 \leq x \leq 360$.

Solution

$2 \cos^2 x° + \cos x° - 1 = 0 \Rightarrow (2 \cos x° - 1)(\cos x° + 1) = 0$

$\Rightarrow \cos x° = \frac{1}{2}$ or -1.

Hence $x = 60$ or 300 $\left(\text{from } \cos x° = \frac{1}{2} \right)$ or $x = 180$ (from $\cos x° = -1$).

It is good form to write the solutions in ascending order, hence $x = 60, 180, 300$.

Hints & tips

We see another reminder of how important it is to be able to factorise a quadratic expression. You could think of the above quadratic equation as $2c^2 + c - 1 = 0$ letting $c = \cos x°$ if it helps. You should also think of the cosine graph to solve $\cos x° = -1$.

Example

Solve the equation $2 \cos^2 x + \cos x - 1 = 0, 0 \leq x \leq 2\pi$.

Solution

The only difference between this equation and the previous one is that we are asked for solutions in the interval $0 \leq x \leq 2\pi$ instead of $0 \leq x \leq 360$. This means that the solution should be given in radian measure. Check that the solution to this equation is $x = \frac{\pi}{3}, \pi, \frac{5\pi}{3}$.

Hints & tips

Always check the interval when solving trigonometric equations to see whether the solution is to be given in degrees or radians.

Example

Solve the equation $2\cos 3x° = 1, 0 \le x \le 360$.

Solution

This example involves a multiple angle $3x°$. This means that when we come to solve for x we will have to divide by 3, therefore we must consider 3 cycles of the graph of $y = \cos 3x°$ which has a period of $120°$ ($360° \div 3$).

$2\cos 3x° = 1 \Rightarrow \cos 3x° = \dfrac{1}{2}$

$\Rightarrow 3x = 60, 300, 420, 660, 780, 1020 \Rightarrow x = 20, 100, 140, 220, 260, 340$

Hints & tips

If you are solving an equation involving a multiple angle, you will probably have to extend the number of cycles. Check that in the above example, after finding that 3x = 60, 300, the further solutions were found by adding 360 (twice) to both 60 and 360.

Identities

A trigonometric identity is an equation that is true for all values of the variables. You may be asked to verify that a trigonometric identity is true. Such proofs are done in a special way, by starting with one side of the equation and manipulating it until you show that it is equal to the other side.

Example

Prove that $\dfrac{\sin^3 x + \sin x \cos^2 x}{\cos x} = \tan x$.

Solution

Start with the left side, looking for a suitable formula to use. If you remember that $\cos^2 x = 1 - \sin^2 x$, then you can substitute this into the left side.

L.S. $= \dfrac{\sin^3 x + \sin x \cos^2 x}{\cos x} = \dfrac{\sin^3 x + \sin x(1 - \sin^2 x)}{\cos x} = \dfrac{\sin^3 x + \sin x - \sin^3 x}{\cos x} = \dfrac{\sin x}{\cos x} =$

$\tan x = $ R.S.

Hints & tips

You could also have started by taking out sin x as a common factor on the numerator. Can you see how to complete the proof in that situation? Never start to prove an identity by writing out the full equation. Always start with one side and prove it equals the other side, as shown above.

For practice

1 What is the exact value of tan 150°?
2 Solve the equation $\sin x° = -\dfrac{1}{\sqrt{2}}, 0 \le x \le 360$.
3 Convert $\dfrac{3}{2}\pi$ radians to degrees.
4 Convert 240° to radians.

What you should know

You should know:

★ the addition formulae for $\sin(A \pm B)$ and $\cos(A \pm B)$
★ the double-angle formulae for $\sin 2A$ and $\cos 2A$
★ how to use these formulae in numerical examples
★ how to use these formulae to solve trigonometric equations
★ how to use these formulae to prove trigonometric identities.

The addition formulae

You should be *very* familiar with the addition formulae in trigonometry. Although you have probably memorised them, they appear in the list of formulae.

$\sin(A + B) = \sin A \cos B + \cos A \sin B$

$\sin(A - B) = \sin A \cos B - \cos A \sin B$

$\cos(A + B) = \cos A \cos B - \sin A \sin B$

$\cos(A - B) = \cos A \cos B + \sin A \sin B$

Example

a) Find an expression equal to $\sin(x + 30)°$.
b) Hence or otherwise find the exact value of $\sin 75°$.

Solution

a) $\sin(x + 30)° = \sin x° \cos 30° + \cos x° \sin 30°$
b) $\sin 75° = \sin(45 + 30)° = \sin 45° \cos 30° + \cos 45° \sin 30°$

$$= \frac{1}{\sqrt{2}} \times \frac{\sqrt{3}}{2} + \frac{1}{\sqrt{2}} \times \frac{1}{2} = \frac{\sqrt{3}+1}{2\sqrt{2}}$$

Hints & tips ⭐

When you are asked to find the exact value of a trigonometric expression, you should not use a calculator. Sometimes you will have to use the table of exact values given in the table in the basic trigonometry chapter so make sure you know them. Check that you understand the simplification of the fraction at the end of the example.

The double-angle formulae

The addition formulae can be adapted to form the 'double-angle' formulae. The following formulae also appear on the list of formulae.

$\sin 2A = 2\sin A \cos A$

$\cos 2A = \cos^2 A - \sin^2 A$

$\cos 2A = 2\cos^2 A - 1$

$\cos 2A = 1 - 2\sin^2 A$

Note that there are three versions of the formula for $\cos 2A$. The second and third versions are frequently used in solving trigonometric equations as you will see shortly. If we change the subject in these two versions to $\cos^2 A$ and $\sin^2 A$ respectively, we form two further formulae which may be needed later if you wish to integrate $\cos^2 A$ or $\sin^2 A$. These are

$\cos^2 A = \dfrac{1}{2}(1 + \cos 2A)$

$\sin^2 A = \dfrac{1}{2}(1 - \cos 2A)$

Keep this in mind in case it is needed for an integration problem.

Using the formulae in numerical examples

Example

If A and B are two acute angles such that $\sin A = \dfrac{4}{5}$ and $\cos B = \dfrac{12}{13}$, find the exact value of $\cos(A - B)$.

Solution

We shall be using the formula for $\cos(A - B)$ but first we must construct two right-angled triangles, one for A and one for B, and use the Theorem of Pythagoras to find the third side in each triangle and hence find the missing ratios.

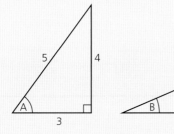

$$\cos(A - B) = \cos A \cos B + \sin A \sin B = \frac{3}{5} \times \frac{12}{13} + \frac{4}{5} \times \frac{5}{13} = \frac{36}{65} + \frac{20}{65} = \frac{56}{65}$$

Hints & tips ★

There was not a problem with signs as both angles were acute. Be careful, however, if one of the ratios is negative; for instance if you are told in an example an angle C was such that $90 < C < 180$, then $\cos C$ would be negative.

You should be aware that you can work back from the formulae to solve some problems.

Example ⚑

In triangle PQR, angle R = 30°. What is the value of $\sin P \cos Q + \cos P \sin Q$?

Solution

You should recognise that $\sin P \cos Q + \cos P \sin Q$ is the right-hand side of the first addition formula and therefore equals $\sin(P + Q)$. In triangle PQR, angles P + Q + R = 180°, hence angles P + Q = 150°.

Hence $\sin P \cos Q + \cos P \sin Q = \sin(P + Q) = \sin 150° = \frac{1}{2}$.

Solving trigonometric equations

Many trigonometric equations can be solved using the double-angle formulae. In these types of equation, the formulae $\cos 2A = 2\cos^2 A - 1$, $\cos 2A = 1 - 2\sin^2 A$ or $\sin 2A = 2 \sin A \cos A$ can be used.

Example ⚑

Solve the equation $2\cos 2x° - 4\cos x° - 1 = 0$, for $0 \le x \le 360$.

Solution

We start by substituting a formula for $\cos 2x°$ into the equation. Choose the formula $\cos 2x° = 2\cos^2 x° - 1$ rather than $\cos 2x° = 1 - 2\sin^2 x°$ because the other term in the equation we have to solve includes $\cos x°$. We will then be able to solve the equation by considering it as a quadratic equation.

$$2 \cos 2x° - 4 \cos x° - 1 = 0$$
$$\Rightarrow 2(2\cos^2 x° - 1) - 4\cos x° - 1 = 0$$
$$\Rightarrow 4\cos^2 x° - 2 - 4\cos x° - 1 = 0$$
$$\Rightarrow 4\cos^2 x° - 4\cos x° - 3 = 0$$
$$\Rightarrow (2\cos x° + 1)(2\cos x° - 3) = 0$$
$$\Rightarrow \cos x° = -\frac{1}{2} \text{ or } \cos x° = \frac{3}{2}$$
$$\Rightarrow x = 120 \text{ or } 240 \ (\cos x° = \frac{3}{2} \text{ is an impossible solution})$$

Hints & tips ★

*A difficult part of the solution occurs when you factorise the quadratic expression $4\cos^2 x° - 4\cos x° - 3 = 0$. If you think of this as $4c^2 - 4c - 3 = 0$ it might help. If you have an **impossible** solution you should say so or you could lose 1 mark. Always check whether the solution has to be given in degrees or radians by inspecting the interval stated in the question.*

Example

Solve the equation $\sin 2\theta = \cos\theta$ for $0 \leq \theta \leq 2\pi$.

Solution

$\sin 2\theta = \cos\theta$

$\Rightarrow 2\sin\theta\cos\theta = \cos\theta$

$\Rightarrow 2\sin\theta\cos\theta - \cos\theta = 0$

$\Rightarrow \cos\theta(2\sin\theta - 1) = 0$

$\Rightarrow \cos\theta = 0$ or $\sin\theta = \dfrac{1}{2}$

$\Rightarrow \theta = \dfrac{\pi}{6}, \dfrac{\pi}{2}, \dfrac{5\pi}{6}, \dfrac{3\pi}{2}$ (it is good form to write solutions in ascending order)

Trigonometric identities

Finally we consider a demanding trigonometric identity. Take your time as you check through each step as we seek to prove that the left side is equal to the right side.

Example

By writing $3x$ as $(2x + x)$ and expanding, prove that
$\sin 3x = 3\sin x - 4\sin^3 x$.

Solution

L. S. $= \sin 3x$

$= \sin(2x + x)$

$= \sin 2x \cos x + \cos 2x \sin x$

$= (2\sin x\cos x)\cos x + (1 - 2\sin^2 x)\sin x$

$= 2\sin x\cos^2 x + \sin x - 2\sin^3 x$

$= 2\sin x(1 - \sin^2 x) + \sin x - 2\sin^3 x$

$= 2\sin x - 2\sin^3 x + \sin x - 2\sin^3 x$

$= 3\sin x - 4\sin^3 x$

$=$ R. S.

For practice

1 By expressing $15°$ as $(60 - 45)°$, find the exact value of $\cos 15°$.
2 If the value of $\sin a° = \dfrac{1}{4}$, where a is an acute angle, find the exact value of $\sin 2a°$.
3 If A and B are acute angles such that $\tan A = \dfrac{3}{4}$ and $\tan B = \dfrac{8}{15}$, find the exact value of $\sin(A + B)$.
4 Solve the equation $\sin 2\theta + \sin\theta = 0$, for $0 \leq \theta \leq 2\pi$.
5 Solve the equation $6\cos 2x° - 5\cos x° + 4 = 0$, for $0 \leq x \leq 360$.

What you should know

You should know:
★ how to solve pairs of equations of the form $\begin{cases} k\cos\alpha = a \\ k\sin\alpha = b \end{cases}$
★ how to express $a\cos x + b\sin x$ in the form $k\cos(x \pm \alpha)$ and $k\sin(x \pm \alpha)$
★ how to sketch the graph of the function $f(x) = a\cos x + b\sin x$
★ how to solve equations of the type $a\cos x + b\sin x = c$.

Solving pairs of equations of the form $\begin{cases} k\cos\alpha = a \\ k\sin\alpha = b \end{cases}$

Example

Solve the pair of equations $\begin{cases} k\cos\alpha° = 6 \\ k\sin\alpha° = -8 \end{cases}$ for $k > 0$ and $0 < \alpha < 360$.

Solution

Start by squaring and adding both equations:

$k^2\cos^2\alpha° + k^2\sin^2\alpha° = 36 + 64 = 100 \Rightarrow k^2(\cos^2\alpha° + \sin^2\alpha°) = 100$
$\Rightarrow k^2 = 100 \Rightarrow k = 10$

(because $\cos^2\alpha° + \sin^2\alpha° = 1$).

Continue by dividing the equations with the equation involving sine on the numerator:

$\dfrac{k\sin\alpha°}{k\cos\alpha°} = \dfrac{-8}{6} \Rightarrow \tan\alpha° = -\dfrac{8}{6}$

We know that cosine is positive (from $k\cos\alpha° = 6$) and sine is negative (from $k\sin\alpha° = -8$), hence α is in the 4th quadrant (using the CAST diagram or 'all, sin, tan, cos').

Hence $\tan\alpha° = -\dfrac{8}{6} \Rightarrow \alpha = 306\cdot9$.

Note that in this type of question $\alpha°$ is called the *auxiliary angle*.

Expressing $a\cos x + b\sin x$ in the form $k\cos(x \pm \alpha)$ and $k\sin(x \pm \alpha)$

The basic sine graph is often referred to as a sine wave and when sine and cosine graphs are stretched, shifted by a phase angle or added they retain their shape. The auxiliary angle can help us to solve many trigonometric problems involving *wave functions* of the type $f(x) = a\cos x + b\sin x$.

To do this we express the wave function as a single function in the form $k\cos(x \pm \alpha)$ or $k\sin(x \pm \alpha)$.

Example

Express $\cos x° - 2\sin x°$ in the form $k\sin(x + \alpha)°$, where $k > 0$ and $0 < \alpha < 360$.

Solution

Start by referring to the list of formulae for $\sin(A + B)$.

Hence $\cos x° - 2\sin x° = k\sin(x + \alpha)°$ becomes

$\cos x° - 2\sin x° = k(\sin x°\cos\alpha° + \cos x°\sin\alpha°) \Rightarrow$

$\cos x° - 2\sin x° = k\sin x°\cos\alpha° + k\cos x°\sin\alpha°$

Now equate coefficients of $\cos x°$ (terms in red) and $\sin x°$ (terms in green) leading to $k\sin\alpha° = 1$ and $k\cos\alpha° = -2$.

Now use the method from the previous section to find k and α.

$k^2 = 1^2 + (-2)^2 = 1 + 4 = 5 \Rightarrow k = \sqrt{5}$ and $\tan\alpha° = \dfrac{1}{-2}$

Check the signs to see that sine is positive and cosine is negative, meaning that α is in the 2nd quadrant. Hence $\tan\alpha° = -\dfrac{1}{2} \Rightarrow \alpha = 153\cdot4$.

We can now say that $\cos x° - 2\sin x° = \sqrt{5}\sin(x + 153\cdot4)°$.

Hints & tips

Once you have reached the stage of $k\sin\alpha° = 1$ and $k\cos\alpha° = -2$ in the above example, the working to find k can be simplified to a Pythagoras-type calculation and it should be simple to get the correct ratio for 'tan', as long as you remember to put the sine coefficient on the numerator and check the signs to find the correct quadrant for α.

Be careful when you are equating coefficients. You may not be able to colour the different parts of the equation, but you could underline one pair.

The graph of the function $f(x) = a\cos x + b\sin x$

Before we can sketch the graph of the function $f(x) = a\cos x + b\sin x$, we should express it as a single function in the form $k\cos(x \pm \alpha)$ or $k\sin(x \pm \alpha)$.

Example

a) Express $\sin x° + 3\cos x°$ in the form $k\cos(x - \alpha)°$ where $k > 0$ and $0 < \alpha < 360$.

b) Find the maximum and minimum values of $\sin x° + 3\cos x°$ and determine the values of x, in the interval $0 \le x \le 360$, at which these maximum and minimum values occur.

$\Rightarrow$

⇨

c) Find where the graph of $y = \sin x° + 3\cos x°$ crosses the y-axis and hence sketch the graph of $y = \sin x° + 3\cos x°$, where $0 \leq x \leq 360$.

Solution

a) $\sin x° + 3\cos x° = k\cos(x - \alpha)° = k\cos x°\cos\alpha° + k\sin x°\sin\alpha°$

Hence $k\sin\alpha° = 1$ and $k\cos\alpha° = 3$.

Hence $k = \sqrt{1^2 + 3^2} = \sqrt{10}$ and $\tan\alpha° = \dfrac{1}{3}$.

As sine and cosine are both positive, $\tan\alpha° = \dfrac{1}{3} \Rightarrow \alpha = 18\cdot4$.

Hence $\sin x° + 3\cos x° = \sqrt{10}\cos(x - 18\cdot4)°$.

b) The maximum value of $\sin x° + 3\cos x° = \sqrt{10}$.

It occurs when $\cos(x - 18\cdot4)° = 1$, that is when $x - 18\cdot4 = 0$ and hence $x = 18\cdot4$.

The minimum value of $\sin x° + 3\cos x° = -\sqrt{10}$.

It occurs when $\cos(x - 18\cdot4)° = -1$, that is when $x - 18\cdot4 = 180$ and hence

$x = 180 + 18\cdot4 = 198\cdot4$.

c) $y = \sin x° + 3\cos x°$ crosses the y-axis when $x = 0$.

This occurs at $y = \sin 0° + 3\cos 0° = 3$, i.e. at $(0, 3)$.

To sketch the graph of $y = \sin x° + 3\cos x°$, sketch the graph of $y = \sqrt{10}\cos(x - 18\cdot4)°$. Think of the graph of $y = \sqrt{10}\cos x°$ moved $18\cdot4°$ to the right.

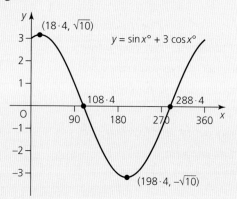

Remember to annotate your graph clearly showing important points such as coordinates of the turning points and points where the graph crosses the x- and y-axes.

Solving equations of the type $a\cos x + b\sin x = c$

We can also solve equations of the type $a\cos x + b\sin x = c$ by expressing $a\cos x + b\sin x$ in the form $k\cos(x \pm \alpha)$ or $k\sin(x \pm \alpha)$.

Example

1 Solve the equation $\sin x° + 3\cos x° = 2$ for $0 < x < 360$.

Solution

$\sin x° + 3\cos x° = 2 \Rightarrow \sqrt{10}\cos(x - 18\cdot4)° = 2$ (from the previous example).

Now divide both sides by $\sqrt{10}$.

Hence $\cos(x - 18\cdot4)° = \dfrac{2}{\sqrt{10}}$. There are two solutions (in the 1st and 4th quadrants as cosine is positive) leading to

$x - 18\cdot4 = 50\cdot8$ or $309\cdot2$. Hence $x = 69\cdot2$ or $327\cdot6$.

Watch out for equations where the solutions are asked for in the interval $0 \le x \le 2\pi$. In this case, there will be no degree symbols and you must work in radians.

2 a) Express $-\cos x - \sin x$ in the form $k\cos(x - \alpha)$ where $k > 0$ and $0 < \alpha < 2\pi$.

   b) Hence solve the equation $-\cos x - \sin x = 1$ for $0 < x < 2\pi$.

Solution

   a) $-\cos x - \sin x = k\cos(x - \alpha) = k\cos x \cos \alpha + k\sin x \sin \alpha$

     Hence $k\sin\alpha = -1$ and $k\cos\alpha = -1$

     $k = \sqrt{(-1)^2 + (-1)^2} = \sqrt{2}$ and $\tan\alpha = \dfrac{-1}{-1} = 1$

     As sine and cosine are both negative, α is in the 3rd quadrant, hence $\tan\alpha = 1 \Rightarrow \alpha = \dfrac{5\pi}{4}$.

     Hence $-\cos x - \sin x = \sqrt{2}\cos\left(x - \dfrac{5\pi}{4}\right)$.

   b) $-\cos x - \sin x = 1 \Rightarrow \sqrt{2}\cos\left(x - \dfrac{5\pi}{4}\right) = 1 \Rightarrow \cos\left(x - \dfrac{5\pi}{4}\right) = \dfrac{1}{\sqrt{2}}$

     As the cosine is positive, there are two solutions (in the 1st and 4th quadrants).

     Hence $\left(x - \dfrac{5\pi}{4}\right) = \dfrac{\pi}{4}$ or $\left(x - \dfrac{5\pi}{4}\right) = 2\pi - \dfrac{\pi}{4} = \dfrac{7\pi}{4}$.

     Hence $x = \dfrac{\pi}{4} + \dfrac{5\pi}{4} = \dfrac{3\pi}{2}$ or $x = \dfrac{7\pi}{4} + \dfrac{5\pi}{4} = \dfrac{12\pi}{4} = 3\pi$.

     Because the second solution is greater than 2π, subtract 2π from it leading to $3\pi - 2\pi = \pi$.

     The solutions to the equation are $x = \pi$ or $x = \dfrac{3\pi}{2}$.

Hints & tips

Some students find it difficult to follow working such as that shown in part (b) because the angles are multiples of π. You can check the working by converting the angles from radians to degrees as you proceed.

In real life situations, the angles which appear may involve multiples of $x°$. Although this makes such situations more complicated, the basic method is the same as before.

Example

Due to tidal variations, the depth of water in a harbour is given by the formula

$D(t) = 1 + \cos 32t° + 2\sin 32t°,$

where $D(t)$ is the depth of water in metres t hours after midnight on a Sunday night.

a) Express $\cos 32t° + 2\sin 32t°$ in the form $k\cos(32t + \alpha)°$ where $k > 0$ and $0 < \alpha < 360$.

b) A boat needs a depth of at least 3 metres of water to leave the harbour. What is the earliest time on Monday that the boat can leave the harbour?

Solution

a) $\cos 32t° + 2\sin 32t° = k\cos(32t + \alpha)° =$

$k\cos 32t°\cos\alpha° - k\sin 32t°\sin\alpha°$

Hence $k\sin\alpha° = -2$ and $k\cos\alpha° = 1$.

$k = \sqrt{(-2)^2 + 1^2} = \sqrt{5}$ and $\tan\alpha° = \frac{-2}{1} = -2$

As sine is negative and cosine is positive, α is in the 4th quadrant, hence $\tan\alpha° = -2 \Rightarrow \alpha = 296 \cdot 6$.

Hence $\cos 32t° + 2\sin 32t° = \sqrt{5}\cos(32t + 296 \cdot 6)°$.

b) Solve the equation $1 + \cos 32t° + 2\sin 32t° = 3$.

Using the solution to part (a), this becomes

$1 + \sqrt{5}\cos(32t + 296 \cdot 6)° = 3$.

$\Rightarrow \sqrt{5}\cos(32t + 296 \cdot 6)° = 2 \Rightarrow \cos(32t + 296 \cdot 6)° = \frac{2}{\sqrt{5}}$.

As cosine is positive, there are two solutions (in the 1st and 4th quadrants).

$\Rightarrow 32t + 296 \cdot 6 = 26 \cdot 57$ or $32t + 296 \cdot 6 = 333 \cdot 43$

$\Rightarrow t = -8 \cdot 43$ or $t = 1 \cdot 15$.

The negative answer is impossible, so the earliest time the boat can leave is $1 \cdot 15$ hours after midnight, that is at 0109 on Monday morning.

For practice

1 Express $3\cos x° - 4\sin x°$ in the form $k\sin(x - \alpha)°$, where $k > 0$ and $0 < \alpha < 360$.

2 Hence solve the equation $3\cos x° - 4\sin x° = 2$ where $0 < x < 360$.

Part 4 Calculus

Chapter 13
Differentiation

What you should know

You should know:
- ★ the basic rules of differentiation
- ★ that the derivative of a function is a measure of the rate of change of the function
- ★ that the derivative of a function is a measure of the gradient of the graph of the function
- ★ how to find the equation of the tangent to a curve
- ★ how to find where a curve is increasing, decreasing or stationary
- ★ how to find the stationary points on a curve and their nature
- ★ how to sketch a curve
- ★ how to find the maximum and minimum values of a function in a closed interval
- ★ how to solve optimisation problems.

Basic rules of differentiation

When we differentiate a function we get the derivative of the function. The notation for the derivative is $f'(x)$ or $\frac{dy}{dx}$ when we differentiate a function with respect to x. Remember the rule:

If $f(x) = ax^n$, then $f'(x) = anx^{n-1}$.

If you are asked to differentiate the sum of different powers of x you should differentiate each term in order. Hence the derivative of the function $f(x) = 3x^3 + 5x^2 - 2x - 6$ is $f'(x) = 9x^2 + 10x - 2$ (note that the derivative of any constant such as 2 is zero). If the function is given in the form $y = 3x^3 + 5x^2 - 2x - 6$ then we say that $\frac{dy}{dx} = 9x^2 + 10x - 2$. It is important that you use a consistent notation for the derivative.

If you meet an example involving brackets, the brackets can be expanded before differentiation, so if you are asked to find the derivative of $(x + 5)^2$, you could say that $(x + 5)^2 = x^2 + 10x + 25$ and then differentiate leading to a derivative of $2x + 10$. Now we shall look at examples involving rational indices, negative indices and surds.

Example

If $f(x) = 4x^{\frac{1}{2}} - 6x^{-\frac{2}{3}}$, find $f'(x)$.

Solution

$f'(x) = 2x^{-\frac{1}{2}} + 4x^{-\frac{5}{3}}$

Hints & tips

Take care when carrying out operations with fractions, especially if this occurs in the non-calculator part of an assessment. Note, for example, that
$-6 \times \left(-\dfrac{2}{3}\right) = 4$. *When calculating* $-\dfrac{2}{3} - 1$, *think of this as* $-\dfrac{2}{3} - \dfrac{3}{3} = -\dfrac{5}{3}$.

Example

1 If $y = 5\sqrt{x} - 2\sqrt[3]{x^4}$, find $\dfrac{dy}{dx}$.

Solution

$y = 5\sqrt{x} - 2\sqrt[3]{x^4} = 5x^{\frac{1}{2}} - 2x^{\frac{4}{3}} \Rightarrow \dfrac{dy}{dx} = \dfrac{5}{2}x^{-\frac{1}{2}} - \dfrac{8}{3}x^{\frac{1}{3}}$

2 $f(x) = \dfrac{3}{x^2} - \dfrac{5}{2x}$. Find $f'(x)$.

Solution

You cannot differentiate while the x terms are on the denominator, so use the laws of indices *before* you start to differentiate.

Hence $f(x) = \dfrac{3}{x^2} - \dfrac{5}{2x} = 3x^{-2} - \dfrac{5}{2}x^{-1} \Rightarrow f'(x) = -6x^{-3} + \dfrac{5}{2}x^{-2} = -\dfrac{6}{x^3} + \dfrac{5}{2x^2}$

3 If $f(x) = \dfrac{3x^2 + 2x - 5}{x}$, find $f'(x)$.

Solution

Before we even start to differentiate we must divide each term on the numerator by x.

$f(x) = \dfrac{3x^2 + 2x - 5}{x} = \dfrac{3x^2}{x} + \dfrac{2x}{x} - \dfrac{5}{x} = 3x + 2 - 5x^{-1}$

Don't think we have finished yet, we have still to differentiate.

Hence $f'(x) = 3 + 5x^{-2} = 3 + \dfrac{5}{x^2}$.

Rate of change

The derivative of a function is a measure of the rate of change of the function. If you are asked to find the rate of change of a function then this question will involve differentiation.

Example

If $A(t) = 2t^2 - 5t + 7$, find the rate of change of A with respect to t when $t = 3$.

Solution

$A'(t) = 4t - 5 \Rightarrow A'(3) = 4 \times 3 - 5 = 7$, hence the rate of change is 7.

The gradient of a curve

The derivative of a function is a measure of the gradient of the graph of the function. Note too that the gradient of a curve at a point on the curve is the gradient of the tangent to the curve at that point.

Example

A curve has equation $y = 2x^3 - 4x$. Find the gradient of the tangent to the curve at the point $(-3, -42)$.

Solution

$\frac{dy}{dx} = 6x^2 - 4$. (Note that this is a formula that can be used to find the gradient at any point on this curve.)

When $x = -3$, $\frac{dy}{dx} = 6 \times (-3)^2 - 4 = 54 - 4 = 50$, hence the required gradient = 50.

Finding the equation of the tangent

Once we know how to find the gradient of the tangent to a curve, we can go further and calculate the equation of the tangent using the formula for the equation of a straight line, $y - b = m(x - a)$.

Example

Find the equation of the tangent to the parabola with equation
$y = x^2 - 8x + 12$ at the point where $x = 2$.

Solution

Find the y-coordinate of the point by substitution. When $x = 2$,
$y = 2^2 - 8 \times 2 + 12 = 0$, hence the point $(2, 0)$ lies on the parabola.
$\frac{dy}{dx} = 2x - 8 = -4$ where $x = 2$, hence the gradient of the tangent $= -4$.
The equation of the tangent is $y - b = m(x - a) \Rightarrow y - 0 = -4(x - 2)$
$\Rightarrow y = -4x + 8$.

Increasing and decreasing functions

We say that a function $f(x)$ is *increasing* if the gradient of the curve is
positive, that is if $f'(x) > 0$.

We say that a function $f(x)$ is *decreasing* if the gradient of the curve is
negative, that is if $f'(x) < 0$.

Example

1 For which values of x is the function $y = x^3 - 12x + 3$ increasing?

Solution

The function is increasing when $\frac{dy}{dx} > 0$, so we start by finding
$\frac{dy}{dx} = 3x^2 - 12$, then solve the inequality $3x^2 - 12 > 0$.

To do this, find the roots of the equation $3x^2 - 12 = 0$ using factorisation.
This leads to $3x^2 - 12 = 3(x^2 - 4) = 3(x + 2)(x - 2) = 0 \Rightarrow x = -2$ or 2.

Next sketch the graph of $y = 3x^2 - 12$. We know that the graph crosses
the x-axis at -2 and 2. We can see the graph must have a minimum
turning point as the x^2 term is positive.

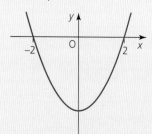

Hence the function is increasing when $x < -2$ or $x > 2$.

2 Explain why the function $f(x) = x^3 + 3x^2 + 3x$ never decreases.

Solution

$f'(x) = 3x^2 + 6x + 3 = 3(x^2 + 2x + 1) = 3(x + 1)^2$
As $3(x + 1)^2 \geq 0$ for all values of x, the function never decreases.

Hints & tips

Problems on increasing and decreasing functions can be solved by differentiation. If you are asked to find the values of x for which a function is increasing and decreasing, you will have to solve an inequality. If the function is cubic, you will have to solve a quadratic inequality.

Stationary points

We say that a function $f(x)$ is *stationary* if the gradient of the curve is zero, that is if $f'(x) = 0$.

At stationary points a function is neither increasing nor decreasing. A stationary point can be a maximum or minimum turning point or a rising or falling point of inflexion.

Maximum turning point	Minimum turning point	Rising point of inflexion	Falling point of inflexion
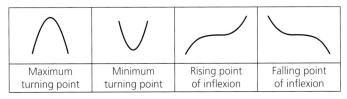			

You must be able to determine which type of stationary point you have found, i.e. the nature of the stationary point.

Example

A function is defined on the set of real numbers by $f(x) = x^3 - 3x^2 + 4$.

Find the coordinates of the stationary points on the curve $y = f(x)$ and determine their nature.

Solution

$f'(x) = 0$ at stationary points (You should write this down as it is worth 1 mark.)

$f(x) = x^3 - 3x^2 + 4 \Rightarrow f'(x) = 3x^2 - 6x = 0$ at stationary points.

$3x^2 - 6x = 0 \Rightarrow 3x(x - 2) \Rightarrow x = 0$ or 2.

Now substitute $x = 0$ and 2 into $f(x) = x^3 - 3x^2 + 4$ to find the y-coordinates of the two stationary points.

When $x = 0$, $f(x) = 4$. When $x = 2$, $f(x) = 2^3 - 3 \times 2^2 + 4 = 8 - 12 + 4 = 0$.

Hence the coordinates of the stationary points are $(0, 4)$ and $(2, 0)$.

We find the nature of the stationary points using a 'nature table' in which we investigate the gradient on either side of the stationary values.

x	$\rightarrow$	0	$\rightarrow$	2	$\rightarrow$
$f'(x) = 3x(x - 2)$	+	0	−	0	+
Slope	/	—	\	—	/

Hence $(0, 4)$ is a maximum turning point and $(2, 0)$ is a minimum turning point.

Hints & tips

When using a nature table, it is recommended that you use the factorised version of $f'(x)$, when possible, to find whether the gradient is positive or negative in the neighbourhood of the stationary values. Check that you follow all the steps above and always set out your working in this way, including a labelled nature table.

Curve sketching

Once we know the stationary points on a curve, we can continue and make a sketch of the curve. Extra information required to do this would be to find where the curve crosses the *x*- and *y*-axes.

Questions on curve sketching require patience as you have to work through several steps before you are in a position to complete the sketch. However, there are many marks available if you succeed. The last example could be part of a longer question in your examination as shown next. The following question could be worth 15 marks in total: 6 for part (a), 5 for part (b) and 4 for part (c).

Example

a) A function is defined on the set of real numbers by
$f(x) = x^3 - 3x^2 + 4$.
Find the coordinates of the stationary points on the curve
$y = f(x)$ and determine their nature.

b) (i) Show that $(x + 1)$ is a factor of $x^3 - 3x^2 + 4$.
(ii) Hence or otherwise factorise $x^3 - 3x^2 + 4$ fully.

c) State the coordinates of the points where the curve with equation $y = f(x)$ meets both the axes and hence sketch the curve.

Solution

a) As shown above, i.e. (0, 4) is a maximum turning point and (2, 0) is a minimum turning point.

b) (i) Use synthetic division.

$$\begin{array}{r|rrrr} -1 & 1 & -3 & 0 & 4 \\ & & -1 & 4 & -4 \\ \hline & 1 & -4 & 4 & 0 \end{array}$$

Hence $(x + 1)$ is a factor of $x^3 - 3x^2 + 4$ as the remainder = 0.

(ii) $x^3 - 3x^2 + 4 = (x + 1)(x^2 - 4x + 4) = (x + 1)(x - 2)(x - 2)$

c) $y = f(x)$ meets the y-axis when $x = 0$, leading to $y = 4$,
so at the point $(0, 4)$
$y = f(x)$ meets the x-axis when $y = 0$, that is $(x + 1)(x - 2)(x - 2) = 0$
leading to $x = -1$ or $x = 2$,
so at $(-1, 0)$ and $(2, 0)$

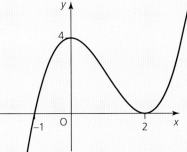

Hints & tips

Note that as the numerical values of x become larger (positive or negative), the 'tails' at either end of the graph continue towards infinity. In your sketch you should show the tails continuing beyond the turning points. Always annotate graphs by showing turning points and points where the graph crosses the x- and y-axes.

It will be useful for you to realise how marks are allocated in extended questions such as the one above. Some of the marks are for communicating what you are doing and can be earned fairly easily. Try to find out how marks are allocated in curve sketching questions. There should be information on marking on the SQA website.

The maximum and minimum values of a function in a closed interval

The maximum and minimum values of a function in a closed interval occur either at a maximum or minimum turning point in the closed interval or at the endpoints of the interval.

Example

Find the maximum and minimum values of the function
$f(x) = x^3 - 3x^2 + 4$ in the closed interval $-1 \cdot 5 \leq x \leq 1 \cdot 5$.

Solution

We can use the sketch of the graph of function $f(x) = x^3 - 3x^2 + 4$ shown above, but taking a snapshot of the graph between the vertical lines $x = -1 \cdot 5$ and $x = 1 \cdot 5$ (the endpoints of the closed interval).

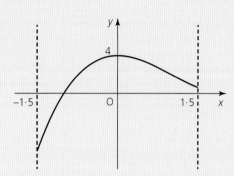

We can see that in this closed interval, the maximum value of the function is 4. It occurs at the highest point of the graph (0, 4) where there is a maximum turning point. The minimum value of the function occurs when $x = -1\cdot5$. This is the lowest point of the graph at the left endpoint. Hence the minimum value is $f(x) = (-1\cdot5)^3 - 3 \times (-1\cdot5)^2 + 4 = -6\cdot125$.

Note that you could do this question *without* drawing a graph by finding there are stationary points at $x = 0$ and $x = 2$ as shown earlier and then evaluating $f(0) = 4$ as this is a turning point. There would be no need to evaluate $f(2)$ as $x = 2$ is outside the closed interval $-1\cdot5 \le x \le 1\cdot5$. Then evaluate $f(-1\cdot5) = -6\cdot125$ and $f(1\cdot5) = 0\cdot625$ (the values at the endpoints). We can see that the maximum value is 4 and the minimum value is $-6\cdot125$.

Optimisation

In optimisation problems, we seek to find the best solution to a practical problem using the techniques we learn in calculus. Problems which ask for maximum/minimum, greatest/least, largest/smallest values for a variable can often be solved by finding stationary values and their nature.

Example

It is planned to include a rectangular room in a leisure centre. The room will contain a rectangular soft play area for young children. The area of the soft play area will be 54 square metres.

The length of the soft play area is x metres.

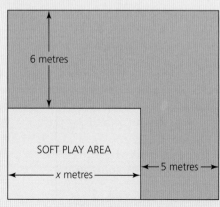

a) If the total area of the room is A square metres, show that
$A = 84 + 6x + \dfrac{270}{x}$.

b) Find the dimensions of the room with minimum area. Justify your answer.

Solution

a) The area of a rectangle is found using $A = lb$.

The length of the room is $(x + 5)$ metres.

As the area of the soft play area is 54 square metres, its breadth is given by $\dfrac{54}{x}$ metres. Hence the breadth of the room is $\left(\dfrac{54}{x} + 6\right)$ metres.

Thus $A = (x + 5)\left(\dfrac{54}{x} + 6\right) = 54 + 6x + \dfrac{270}{x} + 30 = 84 + 6x + \dfrac{270}{x}$.

b) $\dfrac{dA}{dx} = 0$ at stationary points

$A = 84 + 6x + \dfrac{270}{x} = 84 + 6x + 270x^{-1}$

$\Rightarrow \dfrac{dA}{dx} = 6 - 270x^{-2} = 6 - \dfrac{270}{x^2} = 0$ at stationary points.

$6 - \dfrac{270}{x^2} = 0 \Rightarrow 6x^2 = 270 \Rightarrow x^2 = \dfrac{270}{6} = 45 \Rightarrow x = \pm\sqrt{45}$

The length of the room cannot be negative so $x = -\sqrt{45}$ is impossible.

We now use a nature table to verify that the stationary value is a minimum.

x	$\rightarrow$	$\sqrt{45}$	$\rightarrow$
$\dfrac{dA}{dx} = 6 - \dfrac{270}{x^2}$	$-$	0	$+$
Slope	\	—	/

Hence the stationary point is a minimum turning point when $x = \sqrt{45}$. As $\sqrt{45} = 6\cdot71$ approximately, the dimensions of the room with minimum area are:

length $= (x + 5) = (\sqrt{45} + 5) = 11\cdot71$ metres

breadth $= \left(\dfrac{54}{x} + 6\right) = \left(\dfrac{54}{\sqrt{45}} + 6\right) = 14\cdot05$ metres.

Hints & tips ⭐

Optimisation problems often have two parts, the first part being a proof. Even if you cannot complete the proof correctly, use the answer given to do part (b). Remember to include a nature table in your solution, to state if any solutions are impossible and, as we are solving a practical problem, to end the question by clearly stating the solution in words.

The graph of the derived function

You may be given the graph of a function $y = f(x)$ and asked to draw the graph of the derived function $y = f'(x)$. In order to do this successfully, remember the following points:

- a stationary point at (a, b) on $y = f(x)$ means that $(a, 0)$ lies on $y = f'(x)$
- where $y = f(x)$ is *increasing*, the graph of $y = f'(x)$ is positive (above the x-axis)
- where $y = f(x)$ is *decreasing*, the graph of $y = f'(x)$ is negative (below the x-axis)
- the degree of $y = f'(x)$ is one less than the degree of the polynomial $y = f(x)$.

Remember that if the graph of a function is a parabola, then the graph of the derivative is a straight line and if the graph of a function is a cubic, then the graph of the derivative is a parabola.

Example

The diagram shows a sketch of the function $y = f(x)$.

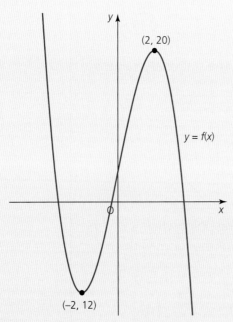

Sketch the graph of $y = f'(x)$.

Solution

The stationary points at $(-2, -12)$ and $(2, 20)$ mean that $(-2, 0)$ and $(2, 0)$ lie on the graph of $y = f'(x)$.

The graph of $y = f(x)$ is increasing (positive gradient) for $-2 < x < 2$. This means that the graph of $y = f'(x)$ is above the x-axis for this interval. It will be below the x-axis for the intervals $x < -2$ and $x > 2$.

The graph of $y = f(x)$ appears to be a cubic function with a maximum and minimum turning point so the graph of $y = f'(x)$ will be a parabola.)

By putting all these facts together we can now draw the graph of the derived function $y = f'(x)$.

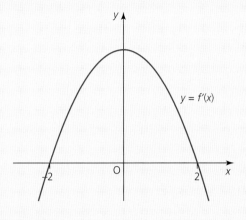

For practice

1 If $f(x) = 2x^3 - 5x + 3$, find the value of $f'(-2)$.

2 If $y = 2\sqrt{x} - 4\sqrt{x^3}$, find $\frac{dy}{dx}$.

3 If $f(x) = \frac{x^2 + 2x + 4}{\sqrt{x}}$, find $f'(x)$.

4 Find the equation of the tangent to the curve $y = (x + 3)^2$ at the point where $x = -1$.

5 If $S(t) = t^2 + 3t - 8$, find the rate of change of S with respect to t when $t = 4$.

6 A function is defined on the set of real numbers by $f(x) = x^3 - 3x + 2$. Find the coordinates of the stationary points on the curve $y = f(x)$ and determine their nature.

7 The total amount of fuel used (in tonnes) by a ship on a voyage of 4000 kilometres is given by the formula $f = 2s + \frac{4000}{s}$ where f is the amount of fuel used and s is the speed in kilometres per hour.

Calculate the speed which gives the greatest fuel economy and the corresponding amount of fuel used for this voyage.

Integration

What is integration?

Integration is the inverse process of differentiation. We know that if we differentiate the function $f(x) = 6x^2 + 4$, we get the derivative $f'(x) = 12x$. To carry out this differentiation, we multiply the $6x^2$ term by the power (2) then subtract 1 from the power. The derivative of the constant term (4) is zero. We can integrate $12x$ with respect to x. The notation is $\int 12x\,dx$. For this inverse process, we add 1 to the power of x and divide by the new power. We must add C for any constant that may have appeared after the variable term. Hence if $f'(x) = 12x$ then $f(x) = \int 12x\,dx = \frac{12}{2}x^2 + C = 6x^2 + C$

Note that C is called the *constant of integration*.

Hints & tips

If you integrate correctly, when you differentiate the solution you will end up with the expression you started with. This type of integral is called an **indefinite integral***. Always include the constant when dealing with indefinite integrals or you will lose 1 mark.*

Basic rules of integration

Remember that $\int ax^n dx = \frac{ax^{n+1}}{n+1} + C$

Example

1 Find $\int 6x^2 dx$.

Solution

$\int 6x^2\,dx = \frac{6x^{2+1}}{2+1} + C = 2x^3 + C.$

$\Rightarrow$

2 Integrate $x(x-5)$ with respect to x.

Solution

We must multiply out the brackets first.

$\int x(x-5)\,dx = \int (x^2 - 5x)\,dx = \frac{1}{3}x^3 - \frac{5}{2}x^2 + C$

3 Find $\int \left(\frac{x^3 + 2x^2 - x}{x} \right) dx$.

Solution

$\int \left(\frac{x^3 + 2x^2 - x}{x} \right) dx = \int \left(\frac{x^3}{x} + \frac{2x^2}{x} - \frac{x}{x} \right) dx = \int (x^2 + 2x - 1)\,dx = \frac{1}{3}x^3 + x^2 - x + C$

Solving equations of the form $\frac{dy}{dx} = f(x)$

Questions in which you are given $\frac{dy}{dx}$ or $f'(x)$ and asked to solve an equation of the form $\frac{dy}{dx} = f(x)$ will involve integration. You could be given extra information about the function $y = f(x)$ to enable you to find the constant of integration, C.

Example

For a certain curve, $\frac{dy}{dx} = 9x^2 - 2x - 4$. The curve passes through the point (2, 4). Find the equation of the curve.

Solution

$\frac{dy}{dx} = 9x^2 - 2x - 4 \Rightarrow y = \int (9x^2 - 2x - 4)\,dx = 3x^3 - x^2 - 4x + C$

Substitute $x = 2$ and $y = 4$ into the equation in order to find C.

$4 = 3 \times 2^3 - 2^2 - 4 \times 2 + C \Rightarrow 4 = 24 - 4 - 8 + C \Rightarrow C = -8$

Hence the equation of the curve is $y = 3x^3 - x^2 - 4x - 8$.

Hints & tips

You should check your solution by differentiation and by substituting x = 2 into the equation.

Example

Find $f(x)$ given that $f'(x) = 2 - \frac{1}{\sqrt[3]{x^2}}$ and $f(8) = 5$.

Solution

$f'(x) = 2 - \frac{1}{\sqrt[3]{x^2}} \Rightarrow f(x) = \int \left(2 - \frac{1}{\sqrt[3]{x^2}} \right) dx = \int \left(2 - \frac{1}{x^{\frac{2}{3}}} \right) dx = \int \left(2 - x^{-\frac{2}{3}} \right) dx$

Hence $f(x) = \int \left(2 - x^{-\frac{2}{3}} \right) dx = 2x - 3x^{\frac{1}{3}} + C = 2x - 3\sqrt[3]{x} + C$.

Substitute $x = 8$ and $f(x) = 5$ into the equation in order to find C.

Hence $5 = 2 \times 8 - 3 \times \sqrt[3]{8} + C \Rightarrow 5 = 16 - 6 + C \Rightarrow C = 5 - 16 + 6 = -5$.

Hence $f(x) = 2x - 3\sqrt[3]{x} - 5$.

Definite integrals

We have looked at indefinite integrals. When we integrate an indefinite integral we get a constant of integration, C. Now we shall study *definite integrals*. A definite integral is one of the type $\int_a^b f(x)\, dx$ where b is the upper limit and a is the lower limit. We evaluate a definite integral using the rule $\int_a^b f(x)\, dx = F(b) - F(a)$. We do not include a constant C when evaluating a definite interval.

Example ▶

Evaluate $\int_1^2 (x^3 - 3x^2)\, dx$.

Solution

$$\int_1^2 (x^3 - 3x^2)\, dx = \left[\frac{1}{4}x^4 - x^3\right]_1^2 = \left[\frac{1}{4} \times 2^4 - 2^3\right] - \left[\frac{1}{4} \times 1^4 - 1^3\right]$$

$$= [4 - 8] - \left[\frac{1}{4} - 1\right]$$

$$= -4 - \left[-\frac{3}{4}\right] = -\frac{16}{4} + \frac{3}{4} = -\frac{13}{4}$$

Hints & tips ★

The above question would probably appear in the non-calculator paper, so you need to be accurate when working with fractions.

The area under a curve

Integration can be used to calculate areas. We can use definite integrals to work out the area under a curve. This refers to the area between a curve and the x-axis. The area may be bounded by the curve, the x-axis and two vertical lines. In general, the area bounded by the curve $y = f(x)$, the x-axis and the vertical lines with equations $x = a$ and $x = b$ where $b > a$ is given by $\int_a^b f(x)\, dx$. If the area found is *above* the x-axis, the area will be positive. If the area found is *below* the x-axis, the area will be negative.

Consider the definite integral worked out in the previous section. We found that $\int_1^2 (x^3 - 3x^2)\, dx = -\frac{13}{4}$. In effect we calculated that the area bounded by the function $f(x) = x^3 - 3x^2$, the x-axis and the lines $x = 1$ and $x = 2$ was $-\frac{13}{4}$ (or $-3\cdot25$). The situation is shown in the graph.

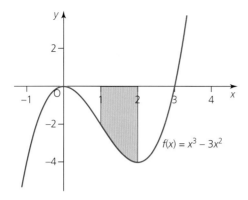

The negative sign in the calculation tells us that the area we have calculated is below the *x*-axis. As we do not consider area to be negative, we remove the negative sign and say that the area bounded by the function $f(x) = x^3 - 3x^2$, the *x*-axis and the lines $x = 1$ and $x = 2$ is $3\cdot25$.

Be careful if you are calculating an area where part of the curve is *above* the *x*-axis and another part is *below* the *x*-axis.

Example

Calculate the total shaded area in the diagram.

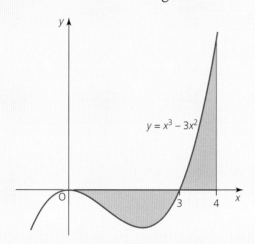

Solution

The curve shown is the same one as in the previous example. However, we must carry out the calculation in two stages, one for the area above the *x*-axis and another for the area below the *x*-axis.

Area below *x*-axis

$$= \int_0^3 (x^3 - 3x^2)\, dx = \left[\frac{1}{4}x^4 - x^3\right]_0^3 = \left[\frac{1}{4} \times 3^4 - 3^3\right] - [0] = \frac{81}{4} - 27 = \frac{81}{4} - \frac{108}{4} = -\frac{27}{4},$$

hence the area is $\frac{27}{4}$.

Area above *x*-axis

$$= \int_3^4 (x^3 - 3x^2)\, dx = \left[\frac{1}{4}x^4 - x^3\right]_3^4 = \left[\frac{1}{4} \times 4^4 - 4^3\right] - \left[\frac{1}{4} \times 3^4 - 3^3\right]$$

$$= [64 - 64] - \left[\frac{81}{4} - 27\right] = 0 - \left[-\frac{27}{4}\right] = \frac{27}{4}, \text{ hence the area is } \frac{27}{4}.$$

Hence the total shaded area $= \frac{27}{4} + \frac{27}{4} = \frac{54}{4} = \frac{27}{2}$ or $13\cdot5$.

Note that if you calculated $\int_0^4 (x^3 - 3x^2)\, dx$ you would get 0 as the areas above and below the x-axis are equal and the positive and negative values would cancel each other out.

The area between two curves

The formula for the area enclosed by the curves $y = f(x)$ and $y = g(x)$ and the lines $x = a$ and $x = b$ for $f(x) \geq g(x)$ and $a \leq x \leq b$ is

$A = \int_a^b \left[f(x) - g(x) \right] dx$.

Example

Calculate the shaded area shown in the graph.

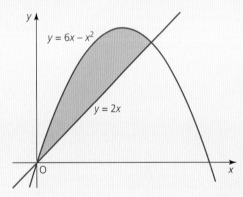

Solution

Before we start to integrate we must find the limits. These will be the x-coordinates of the points of intersection of the curve and the straight line.

To do this solve the equation $6x - x^2 = 2x$.

$6x - x^2 = 2x \Rightarrow 4x - x^2 = 0 \Rightarrow x(4 - x) = 0 \Rightarrow x = 0 \text{ or } 4$

Shaded area $= \int_a^b [f(x) - g(x)]\, dx = \int_0^4 [(6x - x^2) - (2x)]\, dx$

$$= \int_0^4 (4x - x^2)\, dx$$

$$= \left[2x^2 - \frac{1}{3}x^3 \right]_0^4$$

$$= \left[2 \times 4^2 - \frac{1}{3} \times 4^3 \right] - [0]$$

$$= 2 \times 16 - \frac{1}{3} \times 64$$

$$= 32 - \frac{64}{3} = \frac{96}{3} - \frac{64}{3} = \frac{32}{3}$$

Hence shaded area $= \frac{32}{3}$ or $10\frac{2}{3}$.

Hints & tips

When you are calculating the area between two curves, always integrate (the upper curve minus the lower curve) or your solution will be the negative of the correct answer. You do not have to worry whether part of the shaded area is below the x-axis. The formula you use subtracts the area under the lower curve from the area under the upper curve and the solution always works out correctly.

For practice

1 Integrate $3 - \dfrac{8}{x^2}$.

2 Find $\int \left(\dfrac{4x^2 - x}{\sqrt{x}} \right) dx$.

3 For a certain curve, $\dfrac{dy}{dx} = 6x^2 - 2x + 1$. The curve passes through the point $(1, 3)$. Find the equation of the curve.

4 Evaluate $\int_{-2}^{2} (3x - 5)\, dx$.

5 Calculate the total shaded area shown in the diagram.

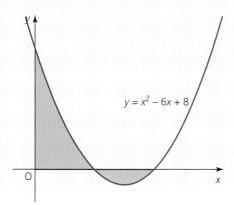

$y = x^2 - 6x + 8$

Further calculus

What you should know

You should know:

★ how to differentiate a function using the chain rule

★ that $\frac{d}{dx}(\sin x) = \cos x$ and $\frac{d}{dx}(\cos x) = -\sin x$

★ that $\int \sin x \, dx = -\cos x + C$ and $\int \cos x \, dx = \sin x + C$

★ that $\int (ax+b)^n \, dx = \frac{(ax+b)^{n+1}}{a(n+1)} + C$

★ that $\int \sin(ax+b) \, dx = -\frac{1}{a}\cos(ax+b) + C$

★ that $\int \cos(ax+b) \, dx = \frac{1}{a}\sin(ax+b) + C$.

The chain rule

The chain rule is used to differentiate *composite functions*. If $y = f(u(x))$, then the chain rule states that $\frac{dy}{dx} = \frac{dy}{du} \times \frac{du}{dx}$. (The first example which follows explains what is meant by a composite function.)

Example

1 If $y = (3x - 2)^6$ find $\frac{dy}{dx}$.

Solution

Consider $(3x - 2)^6$ as being a composition of two functions
$u = 3x - 2$ and $y = u^6$.

Therefore if we are asked to find $\frac{dy}{dx}$ when $y = (3x - 2)^6$,

we can use the chain rule leading to $\frac{dy}{dx} = \frac{dy}{du} \times \frac{du}{dx} = 6u^5 \times 3 =$

$18u^5 = 18(3x - 2)^5$.

In practice, when you are differentiating a bracket raised to a power, multiply the bracket by the power, subtract 1 from the power of the bracket and multiply by the derivative of the contents of the bracket. Check this statement in the above example.

2 If $y = \frac{4}{\sqrt{5x + 6}}$ find $\frac{dy}{dx}$.

Solution

Before we start to differentiate we must put the expression into the form $y = (ax + b)^n$.

Hence $y = \dfrac{4}{\sqrt{5x+6}} = \dfrac{4}{(5x+6)^{\frac{1}{2}}} = 4(5x + 6)^{-\frac{1}{2}}$. Now we can differentiate.

The derivative is $\dfrac{dy}{dx} = 4 \times \left(-\dfrac{1}{2}\right) \times (5x + 6)^{-\frac{3}{2}} \times 5 = -10(5x + 6)^{-\frac{3}{2}} = -\dfrac{10}{\sqrt{(5x+6)^3}}$.

Hints & tips

In the above example the solution was written using a root sign in order to be consistent with the form in which the original expression is given. This is considered to be good form.

The derivatives of sin *x* and cos *x*

When we differentiate sin *x*, we get cos *x*.

When we differentiate cos *x*, we get −sin *x*.

These two key results can be written as $\dfrac{d}{dx}(\sin x) = \cos x$
and $\dfrac{d}{dx}(\cos x) = -\sin x$.

Example

1 If $y = \sin 3x$, find $\dfrac{dy}{dx}$.

Solution

Note that sin 3*x* is a composition of two functions $u = 3x$ and $y = \sin u$, so use the chain rule.

$\dfrac{dy}{dx} = \dfrac{dy}{du} \times \dfrac{du}{dx} = \cos u \times 3 = 3\cos 3x$

In practice, think that the derivative of sin 3*x* is cos 3*x* multiplied by 3 (the derivative of 3*x*).

2 If $y = \cos^3 x$, find $\dfrac{dy}{dx}$.

Solution

Think of $\cos^3 x$ as $(\cos x)^3$, then the derivative of u^3 is $3u^2$ and the derivative of the bracket (cos *x*) is −sin *x*.

Hence if $y = \cos^3 x$, $\dfrac{dy}{dx} = 3\cos^2 x \times (-\sin x) = -3\cos^2 x \sin x$.

A formula for integration

The following formula is very useful when integrating expressions involving brackets:

$$\int (ax+b)^n \, dx = \frac{(ax+b)^{n+1}}{a(n+1)} + C$$

To integrate this type of expression, we reverse the differentiation by adding 1 to the power, and dividing by the (new) increased power. We also multiply the denominator by a, the derivative of the bracket $(ax + b)$.

Example

1 Find $\int (2x+1)^3 \, dx$.

Solution

$$\int (2x+1)^3 \, dx = \frac{(2x+1)^4}{2\times 4} + C = \frac{1}{8}(2x+1)^4 + C$$

2 Evaluate $\int_1^5 \sqrt{2x-1} \, dx$.

Solution

$$\int_1^5 \sqrt{2x-1} \, dx = \int_1^5 (2x-1)^{\frac{1}{2}} \, dx = \left[\frac{(2x-1)^{\frac{3}{2}}}{2\times\left(\frac{3}{2}\right)} \right]_1^5 = \left[\frac{1}{3}(2x-1)^{\frac{3}{2}} \right]_1^5 = \left[\frac{1}{3}\sqrt{(2x-1)^3} \right]_1^5$$

$$= \frac{1}{3}\left[\sqrt{(2\times5-1)^3} - \sqrt{(2\times1-1)^3} \right] = \frac{1}{3}\left(\sqrt{9^3} - \sqrt{1^3} \right) = \frac{26}{3}$$

The integrals of sin $(ax + b)$ and cos $(ax + b)$

There are two very useful integrals which appear on the list of formulae.

$$\int \sin ax \, dx = -\frac{1}{a}\cos ax + C$$

$$\int \cos ax \, dx = \frac{1}{a}\sin ax + C$$

Example

Find $\int \sin 2x \, dx$.

Solution

$$\int \sin 2x \, dx = -\frac{1}{2}\cos 2x + C$$

Hints & tips

There is no excuse for getting this example wrong. The answer is available simply by checking the list of formulae.

However, we can extend the formulae on the list to the two formulae following by replacing ax by $(ax + b)$. The extended formulae are:

$$\int \sin(ax+b)dx = -\frac{1}{a}\cos(ax+b)+C$$

$$\int \cos(ax+b)dx = \frac{1}{a}\sin(ax+b)+C$$

Example

1 Find $\int \sin(4x+3)\,dx$.

Solution

$$\int \sin(4x+3)dx = -\frac{1}{4}\cos(4x+3)+C$$

2 The diagram shows part of the graph of $y = \cos 3x$.

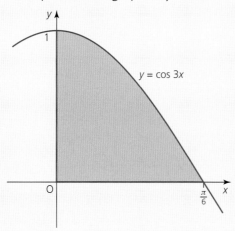

Calculate the shaded area.

Solution

Shaded area

$$= \int_0^{\frac{\pi}{6}} \cos 3x\, dx = \left[\frac{1}{3}\sin 3x\right]_0^{\frac{\pi}{6}} = \frac{1}{3}\left[\sin\left(3\times\frac{\pi}{6}\right)-\sin 0\right] = \frac{1}{3}\times\sin\frac{\pi}{2} = \frac{1}{3}.$$

Hints & tips

If you are working out a definite integral involving a fraction as above, it is a good idea to take out the fraction as a common factor.

Example

Integrate $\cos^2 x$ with respect to x.

Solution

$$\int \cos^2 x\, dx = \int \frac{1}{2}(1+\cos 2x)dx = \frac{1}{2}\int(1+\cos 2x)dx = \frac{1}{2}\left(x+\frac{1}{2}\sin 2x\right)+C$$

This can be simplified to $\frac{1}{2}x+\frac{1}{4}\sin 2x+C$.

Hints & tips

Did you see that $\cos^2 x$ was replaced with $\frac{1}{2}(1 + \cos 2x)$ before we could integrate? This formula is found by changing the subject of the formula $\cos 2x = 2 \cos^2 x - 1$ to $\cos^2 x$ and is mentioned on page 59. Remember that if you are asked to integrate $\cos^2 x$ or $\sin^2 x$, you must change the subject of a formula for $\cos 2x$ before you can proceed.

For practice

1 Differentiate $y = 3 \cos x$.

2 Find $\frac{dy}{dx}$ if $y = (4x - 1)^3$.

3 Find $\int 8 \cos x \, dx$.

4 Evaluate $\int_{-1}^{2} (x + 2)^3 \, dx$.

5 Integrate $\cos(2x - 1)$.

6 Given that $f(x) = \sin 2x + \cos^2 x$ where x is defined on the set of real numbers, find the equation of the tangent to the curve with equation $y = f(x)$ at the point where $x = \frac{\pi}{4}$.

Practice Paper 1

Now you can try a practice exam paper – Paper 1 (<u>non-calculator</u>). Allow 1 hour 30 minutes to do this paper. The standard of questions will be similar to the actual examination and the paper is of the same length. It is worth 70 marks.

There are detailed solutions after the practice paper.

Hints & tips

If a part of a question starts with the phrase 'Hence or otherwise' then you are being told to use the solution to the previous part of the question. The 'otherwise' part of the phrase is to indicate that there is a different (usually more difficult) way of arriving at the solution.

Paper 1 (non-calculator)

1 The points P and Q have coordinates (p, p^2) and $(3q, 9q^2)$ respectively.
   Find the gradient of PQ in its simplest form. **(2)**

2 Find the value of $\int_{-1}^{0} (2x+1)^2 \, dx$. **(4)**

3 Points D, E and F have coordinates (3, 5, 4), (4, 1, 6) and (6, −7, 10) respectively.
   a) Prove that D, E and F are collinear. **(3)**
   b) Find the ratio in which E divides DF. **(1)**

4 Differentiate $y = \frac{(x+2)^2}{2\sqrt{x}}$. **(4)**

5 A straight line has equation $y = 5x - 24$.
   a) Show that this line is a tangent to the parabola with equation
      $y = x^2 - 3x - 8$. **(4)**
   b) Find the coordinates of the point of contact. **(1)**

6 Three lines have equations $3x + 2y - 14 = 0$, $2x - 5y - 3 = 0$ and
   $x + 6y - 10 = 0$.
   Determine whether these lines are concurrent. **(4)**

7 Evaluate $\log_2 6 + \log_2 12 - \log_2 9$. **(3)**

8 a) Express $2x^2 - 8x + 1$ in the form $a(x+b)^2 + c$. **(3)**
   b) State the coordinates of the turning point of the graph
      $y = 2x^2 - 8x + 1$. **(1)**

9 A function is defined on a suitable domain by $g(x) = x^2 + 3$.
  a) Find the inverse function $g^{-1}(x)$.   (2)
  b) State the largest possible domain for $g^{-1}(x)$.   (1)

10 The function f is of the form $f(x) = \log_a(x - b)$.
   The graph of $y = f(x)$ is shown in the diagram.

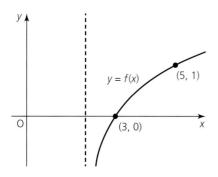

  a) Write down the values of a and b.   (2)
  b) Hence evaluate $f(11)$.   (1)

11 A sequence is defined by the recurrence relation
   $u_{n+1} = 0\!\cdot\!8u_n + 5,$
   $u_0 = 10.$
  a) Calculate the value of u_2.   (1)
  b) Find the limit of the sequence as $n \rightarrow \infty$.   (2)

12 The diagram shows a sketch of the function $y = f(x)$.

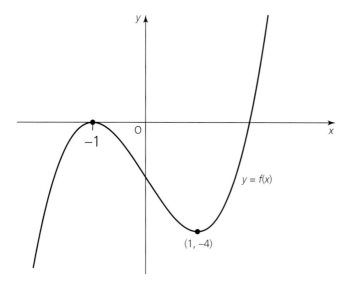

  Sketch the graph of $y = f'(x)$.   (3)

13 Two identical circles touch at the point T $(2, 0)$ as shown in the
    diagram.

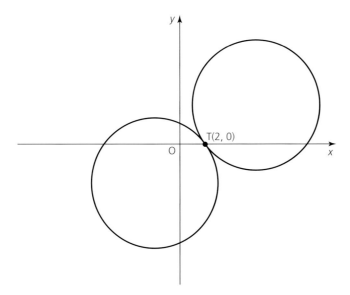

    One of the circles has equation $x^2 + y^2 + 4x + 6y - 12 = 0$.
    Find the equation of the other circle   **(5)**

14 The vectors **u**, **v** and **w** are defined as follows:
        u = 4**i** + **k**,   **v** = 3**i** + 4**j** − 6**k**,   **w** = **j** − 6**k**
    a) Evaluate **u.v** + **u.w**   **(3)**
    b) Hence, make a deduction about the vector **v** + **w**.   **(2)**

15 a) Given that $3x - 1$ is a factor of $6x^3 - 17x^2 + kx + 12$, find the
        value of k.   **(4)**
    b) Factorise the expression fully when k has this value.   **(2)**

16 a) Given that $\dfrac{x}{k} + \dfrac{k+2}{x+1} = 2$, prove that $x^2 + (1 - 2k)x + k^2 = 0$.   **(3)**
    b) Hence find the values of k for which x is real.   **(3)**

17 Two angles A and B are such that
    $(\sin A + \sin B)^2 + (\cos A + \cos B)^2 = 3$.
    Find two possible values for angle $(A - B)$ between 0 and 2π.   **(6)**

Paper 1 – Solutions

1   $m = \dfrac{y_2 - y_1}{x_2 - x_1} = \dfrac{9q^2 - p^2}{3q - p} = \dfrac{(3q - p)(3q + p)}{3q - p} = 3q + p$

2   $\displaystyle\int_{-1}^{0}(2x+1)^2\,dx = \left[\dfrac{1}{2\times3}(2x+1)^3\right]_{-1}^{0} = \dfrac{1}{6}[2\times0+1]^3 - \dfrac{1}{6}[2\times(-1)+1]^3$

    $= \dfrac{1}{6}\times1^3 - \dfrac{1}{6}\times(-1)^3 = \dfrac{1}{6} + \dfrac{1}{6} = \dfrac{2}{6} = \dfrac{1}{3}$

    Note that you could expand the brackets then integrate,

    i.e. $\displaystyle\int_{-1}^{0}(4x^2 + 4x + 1)\,dx$.

3 a) $\overrightarrow{DE} = \mathbf{e} - \mathbf{d} = \begin{pmatrix} 4 \\ 1 \\ 6 \end{pmatrix} - \begin{pmatrix} 3 \\ 5 \\ 4 \end{pmatrix} = \begin{pmatrix} 1 \\ -4 \\ 2 \end{pmatrix}$ and $\overrightarrow{EF} = \mathbf{f} - \mathbf{e} = \begin{pmatrix} 6 \\ -7 \\ 10 \end{pmatrix} - \begin{pmatrix} 4 \\ 1 \\ 6 \end{pmatrix} = \begin{pmatrix} 2 \\ -8 \\ 4 \end{pmatrix}$

Thus $\overrightarrow{DE} = \frac{1}{2}\overrightarrow{EF}$ so DE is parallel to EF with point E in common.

Hence D, E and F are collinear.

b) The ratio in which E divides DF is $1:2$.

4 $\dfrac{(x+2)^2}{2\sqrt{x}} = \dfrac{x^2 + 4x + 4}{2x^{\frac{1}{2}}} = \dfrac{x^2}{2x^{\frac{1}{2}}} + \dfrac{4x}{2x^{\frac{1}{2}}} + \dfrac{4}{2x^{\frac{1}{2}}} = \dfrac{1}{2}x^{\frac{3}{2}} + 2x^{\frac{1}{2}} + 2x^{-\frac{1}{2}}$

Hence the derivative is $\dfrac{3}{4}x^{\frac{1}{2}} + x^{-\frac{1}{2}} - x^{-\frac{3}{2}}$.

5 a) Solve a system of equations to show that $y = 5x - 24$ is a tangent to the parabola $y = x^2 - 3x - 8$.
Hence $5x - 24 = x^2 - 3x - 8 \Rightarrow x^2 - 8x + 16 = 0$
$\Rightarrow (x - 4)(x - 4) = 0 \Rightarrow x = 4$

In the above equation
$b^2 - 4ac = (-8)^2 - 4 \times 1 \times 16 = 64 - 64 = 0.$
$\Rightarrow$ the straight line is a tangent to the parabola.
Alternatively, you could say that the line is a tangent because the roots are equal.

b) When $x = 4$, $y = 5 \times 4 - 24 = -4 \Rightarrow$ point of contact is $(4, -4)$.

6 Find the point of intersection of two of the lines using simultaneous equations. It looks easiest to choose $2x - 5y - 3 = 0$ and $x + 6y - 10 = 0$.

$$2x - 5y - 3 = 0 \qquad\qquad (1)$$
$$x + 6y - 10 = 0 \qquad\qquad (2)$$
(2) × 2: $\qquad 2x + 12y - 20 = 0 \qquad\qquad (3)$

(1) − (3): $\qquad -17y + 17 = 0$
$$\Rightarrow y = 1$$
Substitute $y = 1$ in equation (1): $2x - 5 \times 1 - 3 = 0 \Rightarrow x = 4.$
Hence the lines $2x - 5y - 3 = 0$ and $x + 6y - 10 = 0$ intersect at $(4, 1)$.
Now substitute $(4, 1)$ into the remaining line equation $3x + 2y - 14 = 0$.
Hence $3 \times 4 + 2 \times 1 - 14 = 12 + 2 - 14 = 0.$
Therefore all three lines pass through $(4, 1)$ so they are concurrent.

7 $\log_2 6 + \log_2 12 - \log_2 9 = \log_2 \dfrac{6 \times 12}{9} = \log_2 8 = 3$ (because $2^3 = 8$)

8 a) $2x^2 - 8x + 1 = 2(x^2 - 4x) + 1 = 2(x^2 - 4x + 4) - 8 + 1 = 2(x - 2)^2 - 7$
b) $(2, -7)$

9 a) Let $g(x) = y$, hence $y = x^2 + 3$. Now change the subject of the formula to x.

$$y = x^2 + 3 \Rightarrow x^2 = y - 3 \Rightarrow x = \sqrt{(y - 3)}$$

Hence $g^{-1}(x) = \sqrt{(x - 3)}$ (replace y by x to complete the solution).

b) Because we cannot find the square root of a negative number, $x - 3 \geq 0$. Therefore the largest possible domain is $x \in R, x \geq 3$.

10 a) Substitute $(3, 0)$ into $f(x) = \log_a(x - b) \Rightarrow 0 = \log_a(3 - b)$.
Because $\log 1 = 0$ for all bases, $3 - b = 1$, hence $b = 2$.
Now substitute $(5, 1)$ into $f(x) = \log_a(x - b)$
$\Rightarrow 1 = \log_a(5 - 2) = \log_a 3$.
Because $\log_a a = 1$ for all bases, $a = 3$.

b) $f(11) = \log_3(11 - 2) = \log_3 9 = 2$

11 a) $u_1 = 0{\cdot}8u_0 + 5 \Rightarrow u_1 = 0{\cdot}8 \times 10 + 5 = 13$
$u_2 = 0{\cdot}8u_1 + 5 \Rightarrow u_2 = 0{\cdot}8 \times 13 + 5 = 10{\cdot}4 + 5 = 15{\cdot}4$

b) The sequence has a limit because $-1 < 0{\cdot}8 < 1$.
$L = \dfrac{b}{1-a} = \dfrac{5}{1-0{\cdot}8} = \dfrac{5}{0{\cdot}2} = \dfrac{50}{2} = 25$

12 The stationary points at $(-1, 0)$ and $(1, -4)$ mean that $(-1, 0)$ and
$(1, 0)$ lie on the graph of $y = f'(x)$. The graph of $y = f(x)$ is increasing
(positive gradient) for $x < -1$ and $x > 1$. This means that the graph of
$y = f'(x)$ is above the x-axis for this interval. It will be below the x-axis
for the interval $-1 < x < 1$. The graph of $y = f(x)$ appears to be a cubic
function with a maximum and minimum turning point so the graph
of $y = f'(x)$ will be a parabola.

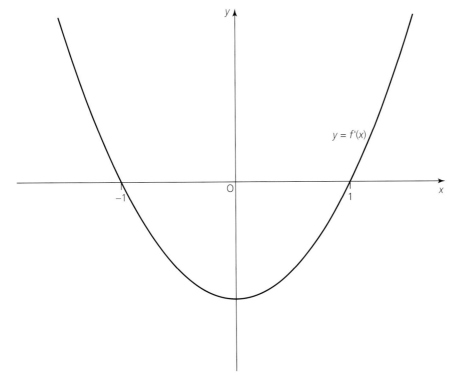

13 The circle with equation $x^2 + y^2 + 4x + 6y - 12 = 0$ has centre $(-2, -3)$.
This is the circle on the left in the diagram.

Its radius is $\sqrt{g^2 + f^2 - c} = \sqrt{2^2 + 3^2 - (-12)} = \sqrt{25} = 5$.

The line joining the centres of the two circles passes through the
point of contact.

The shift from $(-2, -3)$ to $(2, 0)$ is represented by the vector $\begin{pmatrix} 4 \\ 3 \end{pmatrix}$. If we
repeat this shift starting from $(2, 0)$ we find the centre of the circle on
the right in the diagram as the circles are identical.

Hence this circle has centre $(6, 3)$ and radius 5.

Its equation is $(x - 6)^2 + (y - 3)^2 = 25$.

14 a) $\boldsymbol{u}.\boldsymbol{v} = \begin{pmatrix} 4 \\ 0 \\ 1 \end{pmatrix}.\begin{pmatrix} 3 \\ 4 \\ -6 \end{pmatrix} = 4 \times 3 + 0 \times 4 + 1 \times (-6) = 12 + 0 + (-6) = 6$

$\boldsymbol{u}.\boldsymbol{w} = \begin{pmatrix} 4 \\ 0 \\ 1 \end{pmatrix}.\begin{pmatrix} 0 \\ 1 \\ -6 \end{pmatrix} = 4 \times 0 + 0 \times 1 + 1 \times (-6) = 0 + 0 + (-6) = -6$

$\boldsymbol{u}.\boldsymbol{v} + \boldsymbol{u}.\boldsymbol{w} = 6 + (-6) = 0$

b) $\boldsymbol{u}.\boldsymbol{v} + \boldsymbol{u}.\boldsymbol{w} = \boldsymbol{u}.(\boldsymbol{v} + \boldsymbol{w}) = 0$

Hence vector $\boldsymbol{v} + \boldsymbol{w}$ is perpendicular to vector $\boldsymbol{u}$.

15 a)

$$\frac{1}{3} \begin{array}{|ccccc} 6 & -17 & k & 12 \\ & 2 & -5 & \frac{1}{3}(k-5) \\ \hline 6 & -15 & k-5 & 12 + \frac{1}{3}(k-5) \end{array}$$

Hence $12 + \frac{1}{3}(k-5) = 0 \Rightarrow 36 + k - 5 = 0 \Rightarrow k = -31$.

b) $6x^3 - 17x^2 - 31x + 12 = \left(x - \frac{1}{3}\right)(6x^2 - 15x - 36)$
$= (3x - 1)(2x^2 - 5x - 12)$
$= (3x - 1)(2x + 3)(x - 4)$

16 a) Multiply $\frac{x}{k} + \frac{k+2}{x+1} = 2$ throughout by $k(x+1)$.

Hence $\frac{x}{k} + \frac{k+2}{x+1} = 2 \Rightarrow x(x+1) + k(k+2) = 2k(x+1)$.

Hence $x^2 + x + k^2 + 2k = 2kx + 2k \Rightarrow x^2 + x - 2kx + k^2 = 0$.
Hence $x^2 + (1-2k)x + k^2 = 0$.

b) The values of x will be real when the discriminant $b^2 - 4ac \geq 0$.
In this equation $a = 1$, $b = (1-2k)$ and $c = k^2$.
Hence $b^2 - 4ac \geq 0 \Rightarrow (1-2k)^2 - 4k^2 \geq 0$.
Hence $1 - 4k + 4k^2 - 4k^2 \geq 0 \Rightarrow 1 - 4k \geq 0 \Rightarrow -4k \geq -1 \Rightarrow k \leq \frac{1}{4}$.

17 $(\sin A + \sin B)^2 + (\cos A + \cos B)^2 = 3$
$\Rightarrow \sin^2 A + 2\sin A \sin B + \sin^2 B + \cos^2 A + 2\cos A \cos B + \cos^2 B = 3$
$\Rightarrow 2\sin A \sin B + 2\cos A \cos B + 2 = 3$ (since $\sin^2 A + \cos^2 A = 1$)
$\Rightarrow 2(\sin A \sin B + \cos A \cos B) = 1$
$\Rightarrow \sin A \sin B + \cos A \cos B = \frac{1}{2}$
$\Rightarrow \cos(A - B) = \frac{1}{2}$ [from the formula $\cos(A - B) = \cos A \cos B + \sin A \sin B$]
$\Rightarrow A - B = \frac{\pi}{3}$ or $A - B = \frac{5\pi}{3}$

Practice Paper 2

Now you can try another practice exam paper – Paper 2. <u>You may use a calculator in this paper</u>. Allow 1 hour 45 minutes to do this paper. The standard of questions will be similar to the actual examination and the paper is of the same length. It is worth 80 marks.

There are detailed solutions after the practice paper.

Hints & tips

Although you may use a calculator in this paper, in many of the questions it will not be required, so do not become too reliant on your calculator.

Paper 2

1   Triangle XYZ has vertices X $(9, 0)$, Y $(-1, -6)$ and Z $(2, 6)$.
    The median ZM and altitude XA meet at point K.

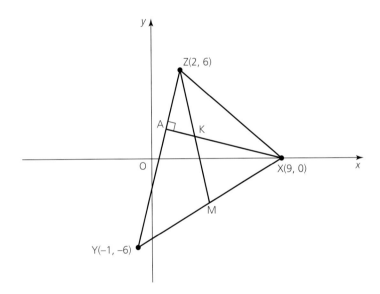

   a) Find the equation of ZM.   (3)
   b) Find the equation of XA.   (3)
   c) Find the coordinates of K.   (3)

2   Solve the equation $2\cos 2x° + 9\sin x° - 4 = 0$, where $0 \le x \le 360$.   (6)

3   A circle has centre C $(6, -4)$.
   a) Find the equation of the circle if it passes through the point A $(0, 5)$.   (2)
   b) Find the equation of the tangent to this circle at A.   (3)
   c) Prove that this tangent passes through the centre of the circle with equation
      $x^2 + y^2 + 12x - 2y + 24 = 0$.   (2)

4 A function is defined on the set of real numbers by $f(x) = x^3 + 2x^2 - 4x - 8$.
   a) Find the coordinates of the stationary points on the curve
      $y = f(x)$ and determine their nature.   **(6)**
   b) (i) Show that $(x + 2)$ is a factor of $x^3 + 2x^2 - 4x - 8$.   **(2)**
      (ii) Hence or otherwise factorise $x^3 + 2x^2 - 4x - 8$ fully.   **(3)**
   c) State the coordinates of the point where the curve with equation
      $y = f(x)$ meets both the axes and hence sketch the curve.   **(4)**

5 Functions f and g are defined on suitable domains by $f(x) = \sin x°$ and
   $g(x) = 2x$.
   a) Find expressions for
      (i) $f(g(x))$ **(1)**
      (ii) $g(f(x))$.   **(1)**
   b) Hence or otherwise prove that
      $g(f(x)) - f(g(x)) = 2\sin x°(1 - \cos x°)$.   **(3)**

6 The diagram shows a cuboid OPQR, STUV.

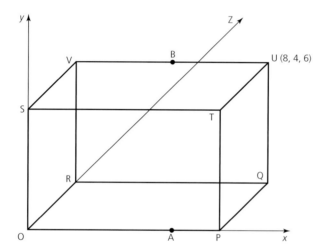

   U is the point (8, 4, 6).
   A divides OP in the ratio 3 : 1.
   B is the midpoint of VU.
   a) State the coordinates of A and B.   **(2)**
   b) Find the components of $\overrightarrow{UA}$ and $\overrightarrow{UB}$.   **(2)**
   c) Find the size of angle AUB.   **(5)**

7 a) Express $12\sin x° - 5\cos x°$ in the form $k\cos(x + \alpha)°$ where
      $k > 0$ and $0 < \alpha < 360$.   **(4)**
   b) Hence solve the equation $12 \sin x° - 5 \cos x° = 10, 0 < x < 360$.   **(4)**

8 A liquid cools according to the law $T_t = T_0 e^{-kt}$ where T_0 is the initial
   temperature and T_t is the temperature after t minutes.

   a) If a liquid cools from a temperature of 100°Celsius to 80°Celsius in
      2 minutes, calculate the value of k.   **(3)**
   b) How long will it take for the liquid to cool to half its original
      temperature?   **(3)**

9 The cost of constructing an oil pipeline is given by the formula

$$C = 128r + \frac{27}{r^2}$$

where C is the cost in millions of pounds and r metres is the radius of the pipeline.

  a) Find the radius which gives the minimum cost of constructing the pipeline. **(6)**

  b) Calculate the corresponding cost of the pipeline. **(1)**

10 The parabola shown in the diagram has equation $y = 25 - x^2$.

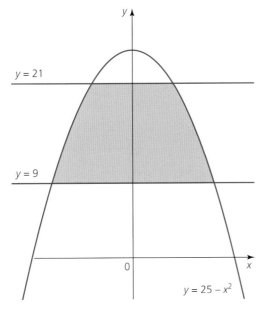

The shaded area lies between the lines $y = 9$ and $y = 21$.
Calculate the shaded area. **(8)**

Paper 2 – Solutions

1 a) The median ZM joins Z to M, the midpoint of XY.

As X is $(9, 0)$ and Y is $(-1, -6)$, M is $\left(\frac{9 + (-1)}{2}, \frac{0 + (-6)}{2} \right) = (4, -3)$.

The gradient of ZM $= \frac{y_2 - y_1}{x_2 - x_1} = \frac{-3 - 6}{4 - 2} = -\frac{9}{2}$.

The equation of ZM is $y - b = m(x - a) \Rightarrow y - 6 = -\frac{9}{2}(x - 2)$.

This simplifies to $2y - 12 = -9x + 18 \Rightarrow 9x + 2y = 30$.

  b) The altitude XA is perpendicular to YZ.

The gradient of YZ $= \frac{y_2 - y_1}{x_2 - x_1} = \frac{6 - (-6)}{2 - (-1)} = \frac{12}{3} = 4$.

Therefore the gradient of XA is $-\frac{1}{4}$.

The equation of XA is $y - b = m(x - a) \Rightarrow y - 0 = -\frac{1}{4}(x - 9)$.

This simplifies to $4y = -x + 9 \Rightarrow x + 4y = 9$.

c) To find the point of intersection of ZM and XA, solve simultaneous equations.

$$9x + 2y = 30 \qquad\qquad (1)$$
$$x + 4y = 9 \qquad\qquad (2)$$
$$(1) \times 2 : 18x + 4y = 60 \qquad\qquad (3)$$
$$(3) - (2) \quad 17x = 51$$
$$\Rightarrow x = 3$$

By substitution in (1): $9 \times 3 + 2y = 30 \Rightarrow y = 1.5$.

Hence the coordinates of K are (3, 1·5).

2 $2\cos 2x° + 9\sin x° - 4 = 0$

Start by replacing $\cos 2x°$ with $1 - 2\sin^2 x°$ (from list of formulae)

Hence $2(1 - 2\sin^2 x°) + 9\sin x° - 4 = 0$

$\Rightarrow 2 - 4\sin^2 x° + 9\sin x° - 4 = 0$

$\Rightarrow 4\sin^2 x° - 9\sin x° + 2 = 0$

$\Rightarrow (4\sin x° - 1)(\sin x° - 2) = 0$

$\Rightarrow \sin x° = \dfrac{1}{4}$ or $\sin x° = 2$

$\Rightarrow x = 14 \cdot 5$ or $x = 165 \cdot 5$

$\sin x° = 2$ is an impossible solution (this *must* be stated).

3 a) Use the distance formula $d = \sqrt{(x_2 - x_1)^2 + (y_2 - y_1)^2}$ to find the radius.

$r^2 = (x_2 - x_1)^2 + (y_2 - y_1)^2 \Rightarrow r^2 = (6 - 0)^2 + (-4 - 5)^2$

$\Rightarrow r^2 = 36 + 81 = 117$

Centre of circle is (6, −4).

To find equation use formula $(x - a)^2 + (y - b)^2 = r^2$.

Hence equation is $(x - 6)^2 + (y + 4)^2 = 117$.

b) The gradient of radius AC $= \dfrac{y_2 - y_1}{x_2 - x_1} = \dfrac{-4 - 5}{6 - 0} = \dfrac{-9}{6} = -\dfrac{3}{2}$.

Hence the gradient of the tangent at A is $\dfrac{2}{3}$.

The equation of the tangent is $y - b = m(x - a) \Rightarrow y - 5 = \dfrac{2}{3}(x - 0)$.

This simplifies to $3y - 15 = 2x$.

c) The centre of the circle with equation $x^2 + y^2 + 12x - 2y + 24 = 0$ is (−6, 1).

Substitute (−6, 1) into

$3y - 15 = 2x \Rightarrow 3 \times 1 - 15 = 2 \times (-6) \Rightarrow -12 = -12$.

Hence the tangent passes through the centre of the circle.

4 a) $f(x) = x^3 + 2x^2 - 4x - 8 \Rightarrow f'(x) = 3x^2 + 4x - 4 = 0$ at SP.

(Note that there is 1 mark for knowing to differentiate and 1 mark for setting the derivative to zero.)

$3x^2 + 4x - 4 = 0 \Rightarrow (3x - 2)(x + 2) = 0 \Rightarrow x = \dfrac{2}{3}$ or $x = -2$

Hence stationary points occur when $x = \dfrac{2}{3}$ or $x = -2$.

When $x = \dfrac{2}{3}$, $f(x) = \left(\dfrac{2}{3}\right)^3 + 2 \times \left(\dfrac{2}{3}\right)^2 - 4 \times \left(\dfrac{2}{3}\right) - 8 = -\dfrac{256}{27}$.

(Use the fraction key on your calculator for the above calculation.)
When $x = -2$, $f(x) = (-2)^3 + 2 \times (-2)^2 - 4 \times (-2) - 8 = 0$.
Nature table:

x	$\rightarrow$	-2	$\rightarrow$	$\frac{2}{3}$	$\rightarrow$
$f'(x) = (3x - 2)(x + 2)$	$+$	0	$-$	0	$+$
Slope	/	___	\	___	/

Hence $(-2, 0)$ is a maximum SP and $\left(\frac{2}{3}, -\frac{256}{27}\right)$ is a minimum SP.

b) (i) Use synthetic division.

$$
\begin{array}{c|cccc}
-2 & 1 & 2 & -4 & -8 \\
   &   & -2 & 0 & 8 \\
\hline
   & 1 & 0 & -4 & 0
\end{array}
$$

Hence $(x + 2)$ is a factor of $x^3 + 2x^2 - 4x - 8$ as the remainder is 0.
(ii) $x^3 + 2x^2 - 4x - 8 = (x + 2)(x^2 - 4) = (x + 2)(x + 2)(x - 2)$

c) The graph of $y = x^3 + 2x^2 - 4x - 8$ meets the y-axis when $x = 0$, that is at $(0, -8)$.
The graph meets the x-axis when $y = 0$. This leads to
$(x + 2)(x + 2)(x - 2) = 0$
$\Rightarrow x = -2$ or $x = 2$ leading to $(-2, 0)$ and $(2, 0)$.

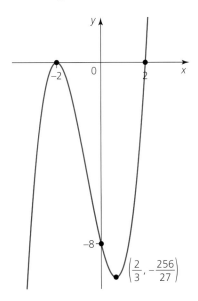

5 a) (i) $f(g(x)) = f(2x) = \sin 2x°$
   (ii) $g(f(x)) = g(\sin x°) = 2\sin x°$
   b) Left side $= g(f(x)) - f(g(x)) = 2\sin x° - \sin 2x°$
   $= 2\sin x° - 2\sin x°\cos x° = 2\sin x°(1 - \cos x°) =$ Right side

6 a) A is $(6, 0, 0)$ and B is $(4, 4, 6)$.

   b) $\overrightarrow{UA} = \mathbf{a} - \mathbf{u} = \begin{pmatrix} 6 \\ 0 \\ 0 \end{pmatrix} - \begin{pmatrix} 8 \\ 4 \\ 6 \end{pmatrix} = \begin{pmatrix} -2 \\ -4 \\ -6 \end{pmatrix}$ and $\overrightarrow{UB} = \mathbf{b} - \mathbf{u} = \begin{pmatrix} 4 \\ 4 \\ 6 \end{pmatrix} - \begin{pmatrix} 8 \\ 4 \\ 6 \end{pmatrix} = \begin{pmatrix} -4 \\ 0 \\ 0 \end{pmatrix}$

c) Use the scalar product that is, $\cos\theta = \dfrac{\mathbf{a.b}}{|\mathbf{a}||\mathbf{b}|}$.

To find angle AUB, we use $\cos\angle AUB = \dfrac{\overrightarrow{UA}.\overrightarrow{UB}}{\left|\overrightarrow{UA}\right|\left|\overrightarrow{UB}\right|}$

$\overrightarrow{UA}.\overrightarrow{UB} = a_1b_1 + a_2b_2 + a_3b_3 = (-2)\times(-4) + (-4)\times 0 + (-6)\times 0 = 8$

$\left|\overrightarrow{UA}\right| = \sqrt{(-2)^2 + (-4)^2 + (-6)^2} = \sqrt{4 + 16 + 36} = \sqrt{56}$

$\left|\overrightarrow{UB}\right| = \sqrt{(-4)^2 + 0^2 + 0^2} = \sqrt{16} = 4$

$\cos\angle AUB = \dfrac{\overrightarrow{UA}.\overrightarrow{UB}}{\left|\overrightarrow{UA}\right|\left|\overrightarrow{UB}\right|} = \dfrac{8}{\sqrt{56}\times 4} \Rightarrow \angle AUB = 74\cdot 5°$

7 a) $12\sin x° - 5\cos x° = k\cos(x + \alpha)° = k\cos x°\cos\alpha° - k\sin x°\sin\alpha°$
Hence $k\sin\alpha° = -12$ and $k\cos\alpha° = -5$.

$k = \sqrt{(-12)^2 + (-5)^2} = \sqrt{169} = 13$ and $\tan\alpha° = \dfrac{-12}{-5}$

As sine and cosine are both negative, α is in the 3rd quadrant.

Hence $\tan\alpha° = \dfrac{12}{5} \Rightarrow \alpha = 180 + 67\cdot 4 \Rightarrow 247\cdot 4$.

Hence $12\sin x° - 5\cos x° = 13\cos(x + 247\cdot 4)°$.

b) $12\sin x° - 5\cos x° = 10$
$\Rightarrow 13\cos(x + 247\cdot 4)° = 10$
$\Rightarrow \cos(x + 247\cdot 4)° = 10/13$
$\Rightarrow x + 247\cdot 4 = 39\cdot 7$ or $320\cdot 3$
$\Rightarrow x = -207\cdot 7$ or $72\cdot 9$
$\Rightarrow x = -207\cdot 7 + 360$ or $72\cdot 9$
$\Rightarrow x = 72\cdot 9$ or $152\cdot 3$

8 a) $T_t = T_0e^{-kt} \Rightarrow 80 = 100\times e^{-2k}$
Now take logarithms to the base e of both sides.
Hence $\ln 80 = \ln(100\times e^{-2k}) = \ln 100 + \ln e^{-2k} = \ln 100 - 2k\ln e$
As $\ln e = 1$, $\ln 80 = \ln 100 - 2k \Rightarrow k = \dfrac{\ln 100 - \ln 80}{2} = 0\cdot 11157$

b) This is a half-life problem. As discussed in Chapter 6, it is a good idea to take the original temperature $T_0 = 1$, so that $T_t = 0\cdot 5$.
Hence $T_t = T_0e^{-kt} \Rightarrow 0\cdot 5 = 1\times e^{-0.11157t} = e^{-0.11157t}$
Again, take logarithms to the base e of both sides.
$\Rightarrow \ln 0\cdot 5 = \ln e^{-0.11157t} = -0\cdot 11157t\ln e = -0.11157t$
Hence $t = \dfrac{\ln 0.5}{-0.11157} = 6\cdot 2$
It takes $6\cdot 2$ minutes for the liquid to cool to half its original temperature.

9 a) At stationary points $\dfrac{dC}{dr} = 0$.

$C = 128r + \dfrac{27}{r^2} = 128r + 27r^{-2}$

$\dfrac{dC}{dr} = 128 - 54r^{-3} = 0$ at SP

Hence $128 - \dfrac{54}{r^3} = 0 \Rightarrow 128r^3 - 54 = 0 \Rightarrow r^3 = \dfrac{54}{128} \Rightarrow r = \sqrt[3]{\dfrac{54}{128}} = 0\cdot 75$.

We now use a nature table to verify that the stationary value is a minimum.

Nature table:

	r	$\rightarrow$	0.75	$\rightarrow$
$\dfrac{dC}{dr} = 128 - \dfrac{54}{r^3}$		$-$	0	$+$
Slope		\	——	/

Hence the stationary point is a minimum turning point when $r = 0.75$.

Therefore a radius of 0.75 metres leads to the minimum cost for the pipeline.

b) $C = 128r + \dfrac{27}{r^2} = 128 \times 0.75 + \dfrac{27}{0.75^2} = 144$.

Hence the corresponding cost of the pipeline is £144 million.

10 To calculate the shaded area we must use integration.

Probably the most efficient method is to calculate the area to the right of the y-axis and double it. We can do this as the y-axis is an axis of symmetry.

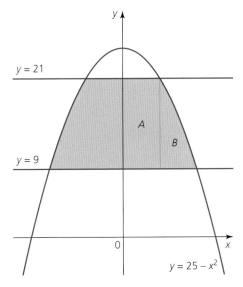

We start by calculating the limits of integration.

$y = 25 - x^2$ meets the line $y = 21$ where

$25 - x^2 = 21 \Rightarrow x^2 = 4 \Rightarrow x = \pm 2$

$y = 25 - x^2$ meets the line $y = 9$ where

$25 - x^2 = 9 \Rightarrow x^2 = 16 \Rightarrow x = \pm 4$

Hence the rectangular shaded area A has area $(21 - 9) \times 2 = 24$.

Area B between $y = 9$ and $= 25 - x^2$

$= \int_2^4 [(25 - x^2) - 9] \, dx = \int_2^4 (16 - x^2) \, dx$.

Hence area B

$= \left[16x - \dfrac{1}{3}x^3 \right]_2^4 = \left[16 \times 4 - \dfrac{1}{3} \times 4^3 \right] - \left[16 \times 2 - \dfrac{1}{3} \times 2^3 \right]$

$= 64 - \dfrac{64}{3} - 32 + \dfrac{8}{3} = \dfrac{40}{3}$

Hence the shaded area $= 2 \times (A + B) = 2 \times \left(\dfrac{40}{3} + 24 \right) = \dfrac{224}{3}$.

Solutions to *For practice* questions

Chapter 1

1 $x^3 - x^2 - 4x + 4$

4 $-0\cdot4, 4\cdot4$

6 $2\sqrt{2}$

2 $3(x+2)(x-4)$

5 $\dfrac{-x-9}{x(x+3)}$

7 $\dfrac{1}{8}$

3 $-\dfrac{2}{3}, 4$

Chapter 2

1 a) Remainder = 0
   b) $(x+1)(x-2)^2$

3 $k = 9$

4 $k = -3; (3x-1)(5x+2)(2x+1)$

2 a) Remainder = 0
   b) $(x-3)(3x+4)(2x-5)$

5 a) $(x+1)(x-1)(x-3)$
   b) $-1, 1, 3$

Chapter 3

1 $x \in R, x \neq -2, x \neq 4$

3 a) $f(g(x)) = x$
   b) They are inverse functions.

2 a) $f(g(x)) = 5x^2 + 9$
   b) $g(f(x)) = 25x^2 - 10x + 3$

4 $f^{-1}(x) = \dfrac{1}{4}(3-x)$

5

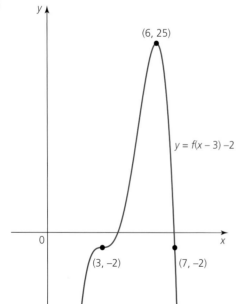

$y = f(x-3) - 2$

(6, 25)

(3, −2)   (7, −2)

Chapter 4

1 $4(x+2)^2 - 19$

2 No real roots

3 $a > \dfrac{25}{16}$

4 $x \le 2$ or $x \ge 3$

5 Proof followed by $(3, 7)$

Chapter 5

1 52

2 $a = 0 \cdot 5, b = 4$

3 $-7 \cdot 5$

4 The litter reaches a limit of 20 kg.

Chapter 6

1 -6

2 $\log_3 10ab$

3 $6 \log_a x$

4 $3 \cdot 024$

5 $x = 2y^3$

6 a) $Y = 0 \cdot 75X + 1 \cdot 5$
   b) $a = 4 \cdot 48$ and $b = 0 \cdot 75$

Chapter 7

1 13

2 $y = 2x - 10$

3 $2x + 3y - 22 = 0$

4 $120°$

5 $3x - 4y - 13 = 0$

6 a) $(4 \cdot 5, 2 \cdot 5)$
   b) $x - 3y + 3 = 0$
   c) $(1 \cdot 5, 1 \cdot 5)$

Chapter 8

1 Centre $(-3, 2)$, Radius $= 6$

2 $(x-4)^2 + (y+1)^2 = 25$

3 $3x - 4y + 25 = 0$

4 a) $(3, 5)$
   b) $x + y = 10$
   c) $(5, 7)$

5 $(x-4)^2 + (y-3)^2 = 25$

Chapter 9

1 $3\sqrt{6}$

2 $\sqrt{21}$

3 $\overrightarrow{AB} = \dfrac{1}{3}\overrightarrow{BC}$ with reason then $1 : 3$

4 $(12, 8, -7)$

5 a) $\overrightarrow{AB} = \begin{pmatrix} -3 \\ 7 \\ -5 \end{pmatrix}$

   b) $\overrightarrow{AC} = \begin{pmatrix} 0 \\ 8 \\ -2 \end{pmatrix}$

   c) $28 \cdot 5°$

Chapter 10

1 $-\dfrac{1}{\sqrt{3}}$ 2 225, 315   3 270°   4 $\dfrac{4\pi}{3}$

Chapter 11

1 $\dfrac{1+\sqrt{3}}{2\sqrt{2}}$

2 $\dfrac{\sqrt{15}}{8}$

3 $\dfrac{77}{85}$

4 $\theta = 0, \dfrac{2\pi}{3}, \pi, \dfrac{4\pi}{3}, 2\pi$

5 $x = 48\cdot 2, 104\cdot 5, 255\cdot 5, 311\cdot 8$

Chapter 12

1 $5\sin(x - 216\cdot 9)°$

2 $13\cdot 3, 240\cdot 4$

Chapter 13

1 19

2 $\dfrac{dy}{dx} = \dfrac{1}{\sqrt{x}} - 6\sqrt{x}$

3 $f'(x) = \dfrac{3}{2}x^{\frac{1}{2}} + x^{-\frac{1}{2}} - 2x^{-\frac{3}{2}}$

4 $y = 4x + 8$

5 11

6 $(-1, 4)$ is a maximum SP; $(1, 0)$ is a minimum SP

7 Speed = $44\cdot 7$ km/h; fuel used = 179 tonnes

Chapter 14

1 $3x + \dfrac{8}{x} + C$

2 $\dfrac{8}{5}x^{\frac{5}{2}} - \dfrac{2}{3}x^{\frac{3}{2}} + C$

3 $y = 2x^3 - x^2 + x + 1$

4 -20

5 8

Chapter 15

1 $-3\sin x$

2 $\dfrac{dy}{dx} = 12(4x - 1)^2$

3 $8\sin x + C$

4 $\dfrac{255}{4}$

5 $\dfrac{1}{2}\sin(2x - 1) + C$

6 $y - \dfrac{3}{2} = -1\left(x - \dfrac{\pi}{4}\right)$